SAS AND ELITE FORCES GUIDE
SNIPER

SAS AND ELITE FORCES GUIDE
SNIPER

SNIPING SKILLS FROM THE WORLD'S ELITE FORCES

MARTIN J. DOUGHERTY

LYONS PRESS
Guilford, Connecticut

An imprint of Globe Pequot Press

Copyright © 2012 Amber Books Ltd
All illustrations © Amber Books Ltd
Published by Amber Books Ltd (www.amberbooks.co.uk)

This Lyons Press edition first published in 2012

Lyons Press is an imprint of Globe Pequot Press.

Library of Congress Cataloging-in-Publication Data is available on file.

ISBN: 978-0-7627-8284-0

Project Editor: Michael Spilling
Design: MRM Graphics
Illustrations: Tony Randell

Printed in Singapore

10 9 8 7 6 5 4 3 2 1

Photos on pages 6–7, 62–63, 176–177, 302–305
courtesy US Department of Defense.

DISCLAIMER
This book is for information purposes only. Readers should be aware of the legal position in
their country of residence before practicing any of the techniques described in this book.
Neither the author nor the publisher can accept responsibility for any loss, injury, or damage
caused as a result of the use of the combat techniques described in this book, nor for any
prosecutions or proceedings brought or instigated against any person or body that may result
from using these techniques.

CONTENTS

A sniper is one of the most highly skilled personnel in any military force. He fires his weapon rarely, but those few shots make a profound difference. Whether protecting friendly forces from enemy marksmen or stalking high-value enemy personnel, the sniper is always vigilant, waiting for the perfect moment to shoot.

PART ONE:
The Sniper

INTRODUCTION

Most people think of snipers as individuals who can shoot with great accuracy. This is certainly true, but there is much more to sniping than marksmanship. Perhaps a better definition of the sniper is a person who has the ability to greatly influence events around him with a single shot.

This in turn assumes a number of things. The sniper must have the ability to get into position to take the shot, and to remain there undetected until it is time to shoot.

Sniper Hides

Concealment lies at the heart of sniping technique. Whether using an elaborate hide or taking advantage of naturally occurring concealment, the sniper must remain unseen while he observes the enemy and sets up his shot.

He must also be able to extract himself safely afterwards. He needs to be able to identify a valuable target and set up a shot that has a good chance of success, taking into account environmental factors like wind and humidity. He must be patient enough to wait for the right moment, and dedicated enough to make the shot without hesitation when the time comes. And then, of course, he must be able to hit the target. There are times when snipers will engage whatever targets present

Target Acquisition

A sniper will get one clear shot at an infantry squad before the survivors take cover. He must select the target that will cause the greatest effect, which will usually be the squad leader. A sniper who is well versed in standard infantry doctrine will know which of these soldiers is in command.

themselves, such as enemy infantrymen. However, this is something of a waste of their abilities. Any soldier can fire at the enemy, and while casualties inflicted on a hostile force will wear down its ability to fight, most casualties are ordinary infantrymen and thus affect the course of an engagement only marginally.

A sniper is trained to select a target whose loss will have a greater effect on the enemy than that of a single infantry soldier. Officers, communications personnel and heavy weapons crews are all high-value targets: their loss degrades the enemy's performance significantly. Eliminating a leader, or depriving him of the ability to give orders, can throw a force into

Heavy Sniper Rifles

RAI Model 500

A 12.7mm (0.5in) bolt-action rifle, the Model 500 is in service with the US Navy and Marines. It is a single-shot weapon and slow to reload as the bolt is entirely removed to chamber each new round. The free-floating barrel is supported by a bipod on the weapon's forearm.

Steyr HS .50

The Steyr HS .50 has a heavy muzzle brake to reduce the recoil generated by its powerful 12.7mm (0.5in) calibre round. A five-shot magazine-fed version is now available.

confusion and prevent it from calling in support from aircraft or artillery.

Higher-level personnel, such as senior commanders or important specialists, are not usually found close to the combat area. A sniper may be able to infiltrate into an area where a conventional force would be quickly detected, and thus has the potential to eliminate key enemy personnel. The loss of a popular commander can demoralize enemy personnel; killing an efficient officer can weaken the whole force.

Snipers can also destroy enemy equipment. Anti-materiel sniping can be used to deprive the enemy of communications or radar equipment, or light vehicles, and weapons can also be targeted. A machine gun whose gunner is disabled can be used by another soldier, but one that is damaged by an armour-piercing round striking the receiver is inoperable in anyone's hands.

Counter-Sniping

The best counter to an enemy sniper is another sniper, and counter-sniper operations are an important part of the sniper's role. A skilled sniper can recognize the places that an enemy might hide by considering where he would position himself; he also

Counter-sniping

Counter-sniper work is one of the most important tasks undertaken by snipers. Often this means passing up other targets in order to remain concealed. Eliminating a skilled enemy sniper is worth letting most other potential victims go.

possesses both the observation skills and the patience to watch for stealthy movement.

He also has the ability to eliminate the enemy sniper from a distance rather than having to move in a patrol to find him. This reduces the exposure of friendly troops to sniper fire and also makes it less likely that the enemy sniper will slip away once the hunt begins to close in.

Law Enforcement Applications

Sniping is often considered to be an exclusively military activity, but there is also an important role to be played in

law enforcement. Military snipers generally shoot from much greater distances than their colleagues in law enforcement, but the principles are, in general, much the same. A law-enforcement sniper may be called upon to disable gunmen who have taken hostages or who otherwise pose an imminent threat to civilians or police officers; but he will usually be required to wait until other, less terminal, means of resolving the situation have been attempted.

This means that sometimes the law-enforcement sniper will have to be ready to shoot at an instant's notice but must wait for a command or for an action on the part of the target that necessitates an immediate shot. This is one difference between military and law-enforcement sniping: a military sniper shoots when the time is right for the shot, while a law-enforcement sniper must often shoot when the circumstances necessitate it. There is no guarantee that the two will coincide.

Law-enforcement snipers also engage in anti-materiel work. The engines of vehicles or boats can be shot out to prevent a suspect from fleeing, or in some cases a weapon can be engaged to render it useless. This requires very precise shooting, as these personnel often operate in an environment where innocent people are nearby, and so must avoid collateral casualties.

1

Sniping is an essential part of modern infantry doctrine, but there have been times when the art was neglected to the point where painful lessons had to be entirely re-learned.

A Brief History of Sniping

The use of projectile weapons to kill or disable a target from a distance predates history. Early hunters learned to throw a spear or shoot a bow to take down prey, and in the case of large or dangerous creatures this required a precise first shot in order to ensure that the prey did not flee or attack the hunter. The same skills were useful in dealing with human threats, such as members of a rival tribe.

spiked club

As organized warfare developed, the emphasis shifted from the use of hunting skills to specialized military techniques, including close-quarters fighting with swords, maces and axes as well as the massed use of archers. The latter would sometimes shoot together in a volley aimed at an area rather than an individual, but they would often take aimed shots at closer range.

Archers honed their skills both in formal practice and in the hunt, and were often extremely proficient marksmen capable of hitting an enemy in the head at 50m (164ft). This was a necessity in an age where enemy infantry were protected by shields and metal body armour. Archers were by no means 'snipers' as such, but their

• •

Snipers have developed from the marksmen of the Eighteenth and Nineteenth Centuries to the highly-trained military professionals of the modern era.

mode of combat called for a high level of marksmanship and some of the skills of a sniper, such as accounting for wind and atmospheric conditions.

Early Firearms

Ironically, as firearms came to dominate the battlefield, the level of marksmanship required of most individual soldiers declined. Early smoothbore matchlock and flintlock firearms were incapable of great accuracy and were most effective when fired in massed volleys at fairly close range. The target was a block of enemy troops, not an individual, and still most shots missed.

Early rifles were slow to load, making them inappropriate for issue to the rank-and-file of most armies. The majority of soldiers received only brief instruction in how to load, point and fire their weapons on command, and few armies spent much on ammunition for live firing practice. Thus there was no point in giving the average soldier a rifle, when he lacked the skills to make much use of it. Instead, he was given a smoothbore weapon, which could make up for lack of accuracy with volume of fire.

Yet even at this low point in the history of marksmanship, there were those who excelled. These individuals developed a high level of skill from hunting in civilian life, and took their marksmanship to war. Whether

Flintlocks and Matchlocks

Accurate shooting with early firearms was a difficult business even without considering the vagaries of smoothbore weapons. Pulling the trigger caused the spring-loaded lock to move forward, scraping the flint against a striking plate or bringing a length of slow-match into contact with gunpowder in the priming pan. Usually (but not always) ignited by the shower of sparks or the slow-match, the powder would hopefully not just 'flash in the pan' but would in turn ignite the main change of gunpowder in the breech. This then burned – slowly compared to modern propellants – and finally began to drive the ball down the barrel. The target or the weapon muzzle could have moved quite a lot in this period, and to make things even harder, the delay from trigger pull to the ball departing the muzzle was not always predictable.

Smoothbore Weapons

Most early firearms such as muskets were muzzle-loaded smoothbores, i.e. they had no rifling to spin the projectile. Firing a spherical ball that was somewhat smaller than the weapon's bore, they were anything but accurate. Muskets were loaded by pouring loose gunpowder down the barrel, then dropping the ball on top. Both were held in place by a paper wad, which was pushed into place using a ramrod.

Smoothbore Balls and Minie Bullets

The hollow base of the Minie bullet expands under the gas pressure of firing to fill the whole barrel and grip the rifling. By contrast, a round ball allows some gas to escape around it and rattles around in the bore, reducing both muzzle velocity and accuracy.

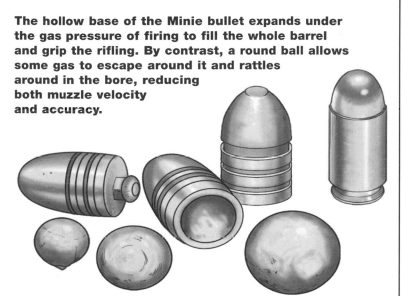

Minie Bullet

The Minie bullet had a conical shape with a hollow base and was slightly smaller than the weapon's bore. When fired, propellant gas pushing the bullet from behind also caused it to expand and grip the rifling. This was an important innovation, as it enabled the creation of fast-firing, rifled, muzzle-loading weapons, which became known as rifle-muskets.

fighting as irregulars or militia, or as part of an organized regiment of troops, these early marksmen could hit individual enemies at distances where a musket shot would be pointless.

As early as the 1770s, the term 'sniper' was in use to describe such individuals, although the context was somewhat different than today. The term originated in the sport of hunting small birds, of which the snipe was one of the hardest to hit. Thus a 'sniper' was an extremely accomplished hunter who had won the respect of his peers, and the term came to be applied in the same context as other words such as sharpshooter or marksman.

When included in a military force, sharpshooters were often, but not always, equipped with rifles instead of smoothbore weapons. They exerted a significant influence on a battle by picking off high-value targets. Sharpshooters could harass enemy artillery by shooting the gunners and could confuse or

Musket Marksmanship

Most muskets did not have sights; there was simply no point in fitting them. Despite this, a good marksman had a reasonable chance of a hit at 100m (328ft) with a well-made musket. At 200–300m (656–984ft), however, the only way to hit a man-sized target was to point the weapon in his general direction and hope to get lucky.

demoralize an enemy formation by killing officers and standard-bearers. They could of course also pick off individuals at considerable range, including their opposite numbers in the enemy force.

Sharpshooters

British troops fighting against the American Revolutionaries (1775–83) came up against such marksmen, who fired from positions of concealment wherever possible. The American sharpshooters favoured long rifles, of which the 'Kentucky Rifle' became the most famous. In fact, the Americans were using several rifle designs at this time, all of which were built for accuracy at ranges far beyond that of a military musket. So, even if

Afghan Sharpshooter

Afghan tribesmen have managed to achieve a surprising degree of accuracy with home-made flintlock weapons called jezails. These weapons outranged the muskets of British infantry, and some remain in use today.

The Kentucky Rifle

Several similar weapons became collectively known as 'Kentucky Rifles'. All were long, black-powder rifles operated by a flintlock mechanism. Although slow to reload, they were accurate out to about 300m (984ft), far further than the smoothbore muskets in use by the military at that time. This made the Kentucky Rifle an ideal hunting weapon that could be highly effective in the hands of irregular marksmen.

they could spot the sharpshooters, troops trained and equipped for massed volley fire were unlikely to be able to return fire effectively.

The British did field their own rifle-armed troops during the American War of Independence, but remained wedded to the concept of massed musket fire and did not create a large rifle corps until 1800, when the Experimental Rifle Corps was created. This formation is better known as the 95th Rifles, as it was named from 1803 onwards. The men of the British rifle corps were equipped with the Baker Rifle, which

was much shorter than the Kentucky Rifles used by the American sharpshooters. It was therefore easier to load and operate under battlefield conditions. The British riflemen were dressed in dark green rather than the red coats that were standard for British infantry at the time.

The choice of clothing for the new rifle force was not about camouflage as such, except in a roundabout way. Uniforms of the period were heavily influenced by traditional dress, and the rifle corps used a uniform inspired by the dark green worn by hunters. Thus the riflemen were given

95th Rifles

The 95th Rifles was an experimental force which proved its worth in the Napoleonic Wars. Despite its success, the rifle was not adopted for general issue, and the Rifles remained something of an elite light infantry force.

a uniform that concealed them from the enemy, which was what they needed, but it was chosen for different reasons.

The rifle troops were considered an elite force and received more training than most infantry forces of the time. In particular, they trained with live ammunition and practised marksmanship, which virtually no unit in any army did at the time. These men were not snipers, however. They were light infantry sharpshooters, often deployed in loose order as skirmishers.

The Baker Rifle

A relatively short, muzzle-loading flintlock rifle, the Baker rifle was issued with a long sword-type bayonet to increase its length for hand-to-hand fighting. The 95th Rifles, which first fielded the Baker Rifle, and units descended from it, used the command 'Fix swords!' rather than 'Fix bayonets!'

Union 1st Sharpshooters

The rifle-muskets of the American Civil War period enabled the individual infantryman to aspire to a high standard of marksmanship. Many 'sharpshooter' units were formed of men who had hunted with firearms all their lives.

Riflemen were trained to operate in pairs, with one man ready to shoot while the other reloaded his weapon. In this way the pair could always respond to a sudden threat or opportunity and could protect one another in a similar fashion to the way modern sniper teams often have a shooter and a spotter who also provides security.

The 95th Rifles were often used like any other infantry regiment, taking part in assaults and fighting alongside regular infantry, but they achieved their best results when permitted to make use of their unique abilities. Taking cover and operating in loose order made the riflemen a difficult target, and they could engage from ranges beyond those of the standard smoothbore musket. This was the weapon of choice of the 95th Rifles' regular opponents, the French *Tirailleurs* (sharpshooters), who were skilled but limited by their weaponry.

Although the 95th Rifles were essentially a conventional light infantry formation equipped with rifles rather than a specialist sniping unit, they did at times perform as snipers in the modern sense. Several generals, as well as innumerable junior officers, in the Napoleonic era fell victim to enemy sharpshooters. One of the most notable was French cavalry General Auguste François-Marie Colbert (1777–1809), who was shot by the legendary rifleman Thomas Plunkett (d. 1851 or 1852). Plunkett lay on his back to make the shot, supporting his rifle barrel with his crossed feet.

Dressed for the Kill

Sharpshooters were also an important part of naval warfare, often positioned high in the rigging of a warship to shoot at enemy officers and marksmen during a close action. It was one such marksman who fatally wounded Admiral Nelson (1758–1805) at the Battle of Trafalgar (1805). Nelson was easily identifiable by his grand uniform and made an attractive target. Although Nelson was shot at the fairly short range of about 15m (49ft), the increasing range of infantry weapons would soon make it desirable for officers to stand out a little less.

Until the mid-1800s, it was common for officers to be dressed ostentatiously to show their status. As accurate long-range weapons became more and more common, ostentation became a death sentence and it became common for officers to dress very similarly to the troops they commanded. During the American Civil War (1861–65), sharpshooters accounted for large numbers of unit commanders and forced a move towards the removal of distinguishing marks.

The practice of infantry officers taking swords into battle became less common during this period, which was partially a survival measure and partly an acknowledgement that a revolver was a better choice of sidearm. Infantry firepower was at that time undergoing a huge leap forward in capability. Most troops were now armed with rifle muskets; black powder weapons, muzzle-loaded like smoothbores but fired using a percussion cap. More importantly, they were rifled.

Confederate Rifleman

The Confederate forces fielded significant numbers of expert marksmen, some of whom were issued with specialist weapons such as the Whitworth rifle. Although primitive by modern standards, the Whitworth gun was deadly in the right hands.

Every Man a Rifleman

With a rifled weapon in every man's hands, both the range and accuracy of infantry firepower increased massively. Smoothbore muskets used a ball that was slightly smaller than the diameter of the muzzle, and were by definition not rifled. A large, often somewhat irregular, lead ball would lose velocity fast and was extremely inaccurate to start with. By using a bullet that was a tight fit in the barrel, and was spun by grooves called rifling, both greater muzzle velocity and accuracy were achieved. This increased not only the range a bullet would travel but, more importantly, the range at which a typical infantryman could reliably hit his target.

The use of rifling was nothing new at this time, but the availability of rifles as standard infantry weapons changed the nature of infantry combat. Accurate personal weapons meant that individual aimed fire was a useful tactic, and of course most infantrymen would choose the most attractive target, which often meant the 'easiest target' – but anyone who stood out was likely to attract shots. Use of cover became increasingly important; close-range firefights in the open gave way to skirmishing tactics using whatever cover and concealment were available, and intense battles over a stone wall or a sunken road became common.

What may be considered sniping tactics in the modern sense began to develop during the American Civil War. Long-range rifles equipped with basic telescopic sights made it possible to shoot accurately over distances of a 1km (0.6 miles) or more. A Confederate marksman killed Union Army General John Sedgwick (1813–64) at a range of more than

Rifling

Rifling consists of grooves cut into the inside of the weapon's barrel, which impart spin and thus stabilize the bullet in flight. To be effective, rifling must be paired with a projectile that fits tightly in the weapon's bore. Ramming a tight-fitting bullet down the barrel of a weapon was a slow process; muzzle-loading rifles fired far more slowly than muskets, which was acceptable when hunting or sniping, but not on the battlefield.

The Whitworth Rifle

Although it was offered to the British military, the Whitworth rifle was not adopted. It became a favourite weapon of Confederate sharpshooters during the American Civil War, and was sometimes used with an early telescopic sight. Although rated as accurate to 800m (875 yards), the Whitworth rifle was used to make a number of shots at significantly longer ranges.

Percussion Caps and Primers

The invention of the percussion cap made firearms far more reliable. Instead of relying on a priming charge of gunpowder ignited by sparks or a slow-match, a weapon could be fired by striking a cap with a spring-loaded hammer, which did away with the need to change flints. This in turn led to the concept of embedding a percussion cap within a cartridge and initiating it with a hammer or pin. Most modern rifle ammunition uses a primer at the base of the cartridge, completing the move from loose powder, ball and wad to an all-in-one cartridge that is both robust and easy to handle.

Enfield Percussion Cap

900m (984 yards), probably using a British-made Whitworth rifle.

Long-range aimed fire rapidly came to dominate the battlefield. From the 1860s onwards, the first European wars were fought using breech-loading rifles. These proved to have massive advantages over muzzle-loaders. Not only could the average infantryman hit an enemy at far greater ranges, but he could reload while lying down, making return fire less effective. This increase in range and accuracy made cavalry attacks all but suicidal and ushered in a period of infantry-dominated warfare.

The Boer Wars

However, it was outside Europe that the real potential of modern rifles was to be demonstrated. In the Boer Wars (1880–81 and 1899–1902), British

'All Quiet Along the Potomac Tonight'

Sniping at the enemy during periods of low activity is nothing new. Even in the mid-Nineteenth Century, snipers were carrying on offensive operations while the bulk of the armies rested and prepared for battle. In 1861, a poem was published in *Harper's Weekly*, in response to a report that all was quiet at the battlefront, except for the shooting of a sentry on picket duty. The first verse goes like this:

"All quiet along the Potomac," they say,
"Except now and then a stray picket
Is shot as he walks on his beat to and fro,
By a rifleman hid in the thicket.
'Tis nothing — a private or two, now and then,
Will not count in the news of the battle;
Not an officer lost — only one of the men,
Moaning out all alone the death rattle."

The poem was first published as 'The Picket Guard' by E. B., later named as Ethel Lyn Beers.

troops came up against highly mobile irregulars equipped with long-range rifles. The Boers were very good marksmen who fought as a large number of snipers rather than a conventional military force. Rather than seek a decisive close-range attack or firefight, the Boers used their skills to pick off British troops at long range. Making excellent use of cover and concealment, the Boers' methods were not very different to those of modern infantry.

In the latter years of the Nineteenth Century, and as a result of experience against enemies such as the Boers, armies moved from colourful uniforms to dowdy and later camouflaged clothing. The intent was to reduce their visibility to enemy

Trench Sniping

troops, making it difficult to pick out a target.

Observation skills not dissimilar to those of a sniper became necessary for battlefield effectiveness, especially once the invention of smokeless gunpowder made it difficult even to see where fire was coming from, let alone target the shooter.

In the more or less static positions of the Western Front, snipers had time to learn where targets were most likely to be found. Low points in the trench parapet, or sections overlooked by higher ground, would expose soldiers to sniper fire as they moved along the trench system.

The Early Twentieth Century

By the early years of the Twentieth Century, industrialization had placed an accurate rifle in the hands of every frontline soldier, and necessity had dressed him in low-visibility clothing. However, this did not by any means make a sniper of every soldier. What set snipers apart from the average soldier was a combination of specialist skills such as target selection and concealment, and a different mindset.

Sniping has more in common with the hunt than with hot-blooded combat. A soldier making or repelling an assault is fighting, to a great extent, for his survival and that of his squad mates. There is an imminent threat to be dealt with, and fighting through to the objective provides a release from danger. In a long-range firefight, he may well be shooting at indistinct shapes glimpsed briefly, and again there will be return fire to provide a very personal level of threat.

For the sniper, it is different. His target is often unaware of him, and may be engaged in some innocent activity. In this the sniper is like the hunter who chooses his victim and kills it because he needs to put food on the table. The sniper has a job to do that must be accomplished without hesitation or regret, but he must do it 'cold'. The majority of people, even trained soldiers, are not capable of this sort of clinical killing.

Live and Let Live

In the trench deadlock of World War I, many soldiers evolved a 'live and let live' attitude to the enemy. They would fight to defend their trench if attacked, and attack if ordered to do so, but would not seek confrontation. A sniper assigned to their sector might be resented by such troops, as he stirred up retribution whenever he took a shot. While others went about the business of survival in the trench, the sniper continued to make war on the enemy.

Snipers in World War I were among the only effective means to attack the enemy without undertaking a massive assault. A sniper, positioned in the trenches or in an elevated position just behind them, could pick off any careless enemy soldier who showed himself. The reason that it is still considered unlucky to be the third person to light a cigarette from the same match is because snipers, in

The Mauser Gewehr 98

A bolt-action rifle produced for the German Army at the end of the Nineteenth Century, the standard Gewehr 98 was robust and accurate. At the outbreak of World War I, the best examples from the production line were adopted for sniper work, and fitted with a telescopic sight. Shortened Gewehr 98s armed German infantry during World War II, while the sniper version continued to give good service. Several successful rifles have drawn heavily on the original Mauser design.

this period, would spot the match being struck in the dark and would be ready to shoot as the third smoker stepped up for a light.

The sniper would watch a spot on the enemy parapet or a firing port made in a trench wall, waiting for an enemy soldier to show himself either to move or fire. He might wait for hours or even days to make a single shot. Enemy equipment was also attacked. Snipers were issued armour-piercing ammunition to shoot the breech-blocks out of enemy machine guns. They would also shoot at the 'trench periscopes' used to peer over the parapet in safety.

Observation and Marksmanship

Sniping in World War I generally made use of prepared positions rather than a long, stealthy move into a firing position, and it was observation and marksmanship that were the key skills at the time. This was especially important when trying to eliminate an enemy sniper who might present a very small target when firing through a tiny gap in the trench parapet. However, snipers did at times sneak out into no-man's land in order to find a good position to shoot from. This was extremely dangerous and required excellent concealment. Detection meant being trapped with little cover and no way to return to friendly trenches.

The German Army had a well-evolved sniper doctrine at the

outbreak of World War I, but it was some time before the Allies caught up. The British Army did not implement seemingly obvious ploys, such as watching gaps or low points in the enemy parapet and picking off soldiers as they passed, possibly because it may have seemed somehow 'unsporting'. However, by 1916 the British Army had gained a specialist training facility for dealing with sniping, observation and scouting – an acknowledgment that, for the sniper, these skills are inextricably intertwined.

The Ghillie Suit

It was around this time that the science of camouflage began to be developed. Early disruptive-pattern materials were created experimentally, and snipers operating in areas where not all the vegetation had been killed by the conflict began to attach foliage to their clothing as concealment. These two measures did more than simply blend in with the surroundings: they also broke up the characteristic lines of man and rifle, making the sniper much harder to recognize as a human being rather than part of the landscape.

This concept was taken further by the introduction of the ghillie suit, developed by Scottish gamekeepers known as ghillies. The ghillie suit was made of strips of soft cloth, to which vegetation could be added. A sniper could thus alter his suit to better

Early Ghillie Suit

Although by no means as elaborate as a modern ghillie suit, this World War II sniper's hood and coat will still break up the characteristic outline of his body. Most importantly, his head is changed from an easily recognisable shape to a soft, shapeless mass.

match local conditions. The ghillie suit turned a sniper into a shapeless mass that was much the same colour as the background, although its effectiveness was reduced if the straight lines of the rifle were not also broken up.

World War II

Snipers proved their worth as both combat and observation assets during World War I, but despite the lessons learned, sniping was not greatly emphasized in most nations between the wars. It was not until another major war broke out that snipers began once more to play an integral role in combat.

There is no doubt that snipers were active during the Spanish Civil War of the mid- to late 1930s (although they were more likely to be concealed marksmen than formally trained experts), but it was not until Russia invaded Finland in 1939 that sniping began to attract international attention. The vastly outmatched Finnish Army managed to fight the Red Army to a standstill, making use of the arctic forest terrain in which much of the fighting took place.

In addition to supporting conventional troops, Finnish snipers slipped behind Russian lines and inflicted demoralizing casualties on

Skin Camouflage

Patterns of colour on the face can disguise the features, making it far less likely that an observer will resolve what he sees as a face. This in turn makes it less likely that the sniper will be spotted; the face and head are among the most obvious giveaways.

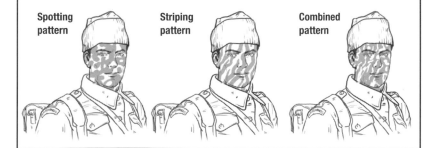

Spotting pattern

Striping pattern

Combined pattern

Simo Häyhä (the 'White Death') (1905–2002)

Simo Häyhä was a marksman in the Finnish Army when his country was invaded by Russia in 1939. Normally armed with a bolt-action rifle, Häyhä made many of his 700 or more kills with a sub-machine gun. At times, he used his rifle's iron sights rather than a telescopic sight to lower his profile, and would hold snow in his mouth to prevent his breath from condensing and giving away his position. Although he was shot in the face by a Russian sniper, Häyhä survived the war and retired as a national hero.

forces who felt that they were safe. This required masterful use of camouflage and concealment, and one Finnish sniper emerged as the greatest of all time. This was Simo Häyhä (1905–2002), who became known as the 'White Death'. He killed around 700 enemy troops, often using iron sights (the standard sights supplied on a rifle) rather than a telescopic sight. This lowered the profile of his rifle and made detection even harder.

Counter-snipers

After the outbreak of World War II, armies of all nations set up sniping schools or expanded existing ones. Snipers were provided as roving support for conventional forces on the offensive, stalking and eliminating enemies who were holding up an advance. This included other snipers but also the crews of well-positioned anti-tank guns and support weapons.

Counter-sniper work became extremely important as the number of snipers on all sides increased. A single sniper could weaken the morale of troops who were resting or hold up an advance for an excessive amount of time. The knowledge that there might be a marksman out there seeking a target could make troops unwilling to leave cover – men who would brave intense but impersonal machine-gun fire were unsettled by the idea of a sniper who would single them out to be killed. The precise, deliberate and personal nature of sniper fire was as demoralizing as it was effective, and had to be countered quickly. The best

Scrape OP

A scrape is essentially a shallow trench. Although it will provide some protection from enemy fire if necessary, the scrape is primarily a concealed position. It is camouflaged above with both artificial and natural materials such as vegetation from the local area.

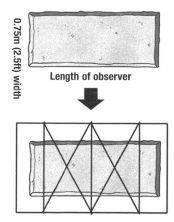

0.75m (2.5ft) width

Length of observer

- Lateral rope stakes
- Covering emplaced and pegged

- Camouflage netting
- Topsoil and camouflage added

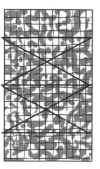

Kar 98 Sniper Rifle

The sniper variant of the Karabiner 98 was not produced as a specialist sniping weapon. Instead, the most accurate examples were selected by testing every rifle that came off the production line.

response was usually to send counter-snipers to find and eliminate the enemy marksman.

The very best snipers were sometimes assigned special missions to eliminate a particular target. This might be a prominent enemy officer, but often was a highly successful sniper. Some of these 'super snipers' turned out not to be a single individual but a group of snipers who had enjoyed a run of success. However, at times it was a single individual who had inflicted heavy losses or made it impossible to enter a particular area. Sniper duels were, of necessity, drawn-out affairs involving long days of observation and analysis of the opponent's methods. If the game of concealment

versus observation were not won, the counter-sniper would become another victim. Thus the shot that ended a sniper duel was merely the endgame in a very long contest.

Urban Combat

World War II was characterized by, among other things, urban combat. Not only were great cities fought over and through but any village or town could be a strongpoint for enemy troops. Tanks were vulnerable in urban terrain, so clearing a town was an infantry task. This created ideal conditions for snipers, and so a new body of technique, urban sniping, was created. The principles were much the same but some particulars had to be adapted.

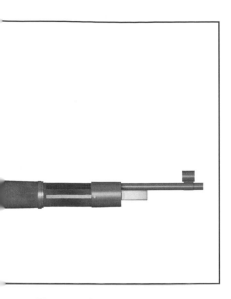

The most famous of all urban combats in World War II was the battle for Stalingrad. The wreckage of the city became a deadly hunting ground for patrols and snipers. In between large-scale battles there were endless skirmishes between small units, which were often unable to offer mutual support due to the complex terrain. It soon became an accepted fact that to emerge from a building was likely to be fatal, so soldiers broke through interior walls and moved under cover wherever possible.

In such an environment, a lone sniper could deny a sizable area of the city to the enemy, and the assumption of his presence would impede operations even after he had moved to other hunting grounds. The most feared Russian sniper of the period was Vassili Zaitsev (1915–91), who became a hero among his own forces after killing over 100 German soldiers. In an effort to neutralize Zaitsev and others like him, experienced German snipers were deployed to Stalingrad.

This has been portrayed as a duel between Zaitsev and his opposite number on the German side, who may have been a sniper instructor named Major Erwin Koenig (whose existence is disputed). However, it is just as likely that what occurred was a concerted effort to deal with Russian snipers in general, or even just a natural consequence of the fighting in Stalingrad. In other words, rather than the German Army deploying a single 'super sniper' to hunt down Zaitsev, it is possible that both sides simply engaged in more general counter-sniper operations, which naturally involved Zaitsev.

Zaitsev became aware of an extremely skilled German sniper – who may have been Koenig, and may or may not have been sent to Stalingrad to hunt him down – when some of his highly competent colleagues were killed. By a combination of observation and deduction about where he himself would hide, Zaitsev located the enemy sniper's hide and lured him into exposing his position. Zaitsev then killed his opponent with a single

Soviet Sniper Team

The Red Army made extensive use of snipers during World War II, having learned their value during the conflict with Finland.

Soviet snipers generally used the Mosin-Nagant rifle, though some preferred semi-automatic weapons. Red Army snipers were generally equipped with white smocks for winter operations.

Vassili Zaitsev (1915–91)

Vassili Zaitsev is particularly associated with the Battle of Stalingrad, one of the turning points of World War II. As the battle stagnated into bitter urban fighting, Zaitsev killed over 100 German troops, including several snipers. It is thought that the German Army was so concerned about this 'super sniper' that they brought in one (or possibly several) of their best to hunt him down. According to one version of the story, Zaitsev became aware that he was being hunted after a sniper killed some of his highly skilled comrades. He and his opponent stalked one another for several days before Zaitsev worked out where the German's hide was located and shot him dead. Zaitsev himself survived the battle and the rest of the war.

shot. The incident became something of a legend, although nobody is sure of all the facts.

Transferable Skills

The Red Army understood the value of passing on the knowledge that experts had gained in the field, and was willing to pull snipers out of the field to train others. One such was Lyudmila Pavlichenko (1916–74). One of a large number of female snipers fielded by the Red Army, Pavlichenko accounted for 36 enemy snipers and nearly 300 other personnel before being transferred to a sniper school to pass on her skills. However much damage Pavlichenko might have

been capable of doing to the enemy, her value as a propaganda asset was arguably greater, and the skills she taught to her hundreds of pupils greatly increased their effectiveness and life expectancy. Although none of her pupils achieved as much as she might have, Pavlichenko continued to inflict massive damage on the German war effort through them.

Snipers in the Pacific

Snipers were also highly active in the Pacific theatre of war. Japanese snipers were adept at concealing themselves in the foliage high up in a tree, which not only gave them an elevated shooting position but also

The Mosin-Nagant M1891 Rifle

Chambered for 7.62 x 54mm (0.3in) ammunition, the Mosin-Nagant was a clip-loaded, bolt-action rifle with a five-round internal magazine. Like other weapons of its generation, it represented a great leap forward in infantry firepower compared to earlier single-shot breech-loaders. A shortened variant of the infantry version served in World War II, while the Mosin-Nagant sniper version was also used in the Korean and Vietnam conflicts.

placed them above the normal eye line of troops operating on the ground. The tree provided both camouflage and some cover if the sniper were spotted, although it also made a quick retirement problematical.

US forces learned to saturate likely sniper positions with automatic weapons fire in the hope of a lucky hit. A sniper whose locality was receiving a huge volume of fire was unlikely to be able to make an accurate shot, so area suppression was a reasonable counter if the target could not be identified. A more

effective variation on this method was to shoot at likely sniper hides in the hope that the suspected sniper would move or return fire. His position was then smothered by intense fire.

Anti-tank rifles, mortars, light artillery and anti-tank guns were all used to counter Japanese snipers – as well as infantry weapons and machine guns. The thick branches of jungle trees might protect the sniper from lighter rounds but a powerful anti-tank rifle could penetrate even quite dense protection. Heavy fire would also strip away foliage and

Lyudmila Pavlichenko (1916–74)

The Soviet Union fielded large numbers of female snipers in World War II, many of whom were trained by Lyudmila Pavlichenko. Pavlichenko was already a skilled marksman and tried to join the army when war broke out, but had to overcome the assumption that as a woman she was better suited to becoming an army nurse than a combat soldier. Eventually, she was able to join a sniper unit, and was hesitant about killing until she witnessed the death of a nearby comrade. After that, she was entirely willing to shoot at the enemy and became an effective sniper. Normally using a semi-automatic SVT-40 rifle, Pavlichenko established a fearsome reputation in the fighting around Odessa and Sevastopol (1941–42), after which she became an instructor. She survived the war and was honoured as a Hero of the Soviet Union. Her husband, also a sniper, was killed in action at Sevastopol.

Type 99 Rifle

The Japanese armed forces adopted the Type 38, a short, light bolt-action rifle chambered for a 6.5mm (0.25in) cartridge for general use, but found it to be underpowered.

make the position too exposed for use, so sometimes this was done in preparation for an advance. If enemy snipers were present, they might be hit. If they were not, it was unlikely that they would try to move into an area without adequate concealment.

It was customary at the time to use a two-man sniper team, consisting of a sniper and an observer, who might be a less experienced sniper receiving advanced instruction from a master. Other teams consisted of two 'shooters' who would swap roles, or a long-term partnership between a sniper and a skilled observer. The US Marine Corps took a different approach, fielding sniper teams of three men.

Marine sniper teams added a third man for security. While the sniper and observer searched for targets, the third man covered their backs. Rather than a sniping weapon he carried an infantry rifle, a sub-machine gun or perhaps a shotgun. If a close-range fight developed, sniper weapons were limited in their effectiveness and the third man's ability to lay down intense fire would allow the team to break contact or fight clear of a difficult situation.

Postwar Sniping

As at the end of the previous major conflict, sniping was neglected in the period immediately after the end of World War II. As a result, the troops sent to South Korea in response to the invasion by North Korea in June 1950 were forced to re-learn all the lessons their predecessors had absorbed.

This was often on an ad hoc basis, leading to some interesting improvisations. Ordnance Captain

The Type 99 used a more powerful 7.7mm (0.303in) round. This proved to be an effective sniping weapon when fitted with a telescopic sight. Japanese sights were fixed; i.e. they could not be adjusted for range. Snipers were expected to compensate for bullet drop by manually adjusting their aim point. Despite this disadvantage they proved deadly in the Pacific War.

William Brophy, a US sniper who was also a gunsmith, came up with the idea of using a modified 0.5in (12.7mm) calibre heavy machine gun as a sniping weapon from static positions. This weapon was too bulky for mobility in the field, so Brophy converted an anti-tank rifle to use a 0.5in calibre machine-gun barrel and chamber, creating the forerunner of today's heavy anti-materiel rifles.

The Korean War eventually settled down into a static confrontation, with heavily defended hills forming the main positions on both sides. In between set-piece assaults and ferocious close-range fighting to try to break through the enemy lines, sniping was a constant harassment. Communist snipers used tunnels to move into position unobserved, and learned to make use of landmarks,

US Marine Sniper, 1951

Although the Springfield M1903 was intended to be phased out as a general-issue infantry rifle in favour of the M1 Garand, large numbers were retained as sufficient M1s were not always available. Most sniper candidates were thus already familiar with the weapons they were to be issued.

improving their accuracy as they essentially made the same shot over and over again.

US troops came up with some innovative ideas. One was for a sniper to be supported by several automatic weapons. Lacking telescopic sights, the gunners were unlikely to be able to spot a target, but they could fire at the general area indicated by a sniper using tracer ammunition. Even if the sniper did not hit his target, he could make the general area extremely unhealthy to inhabit for a while, possibly inflicting casualties or, at least, bringing down accurate suppressing fire.

Snipers were also used to support an attack. Communist troops were wary of raising their heads above their defences under most conditions, but when an assault force approached, they had to rise up to fire. This provided US snipers with many targets that were stationed back from the attack and not under fire themselves, which they could pick off at leisure.

Vietnam
US involvement in the Vietnam War (1965–73) pitted a technologically sophisticated force against insurgents who often operated in small groups and dispersed in the face of significant enemy forces. The North Vietnamese and Viet Cong made extensive use of snipers to harass US bases and patrols, which

necessitated a robust anti-sniper doctrine. One technique was to saturate possible hiding places with heavy firepower, a practice that, although wasteful, could reduce the danger from concealed enemy marksmen. US snipers played an important part in counter-sniper operations, and also in a wider context. The US Marine Corps explicitly trained its snipers also to act as scouts, making use of their concealment and observation skills to gather information on enemy movements or enable friendly patrols to avoid an ambush.

Suppressors
US forces discovered that although the M14 rifle was not a great success as a battle rifle, it could be the basis for an excellent sniping weapon. This gave rise to the M21 semi-automatic sniper rifle, which could be fitted with a suppressor. This enabled snipers to make several rapid shots if necessary, and to remain concealed even when firing from close to enemy forces. Although a suppressor did not make a weapon 'silent', it did make its sound less likely to be identified as a gunshot and the origin harder to pinpoint. The sound of gunfire can normally be heard a long way off; a suppressor reduced the range at which a sniper's weapon would attract attention. The M21 could also be used with a low-light sight, enabling accurate shooting even at night.

M21 Sniper Rifle

The M14 was not a success as a battle rifle, but an accurised version proved very effective as a sniper weapon and was adopted into service as the M21. Effective to about 700m (766 yards), the M21 could take a suppressor and a variety of specialist optical devices as well as a normal telescopic sight.

US snipers were taught to make use of all available assets. By shooting an enemy patrol's officer, and then anyone who seemed to be showing any initiative, they could pin down a quite large force for some time. This situation could be exploited further by then calling in air support or artillery fire on the enemy unit before slipping away.

Sergeant Carlos Hathcock

In one of the most famous exploits of this type, Sergeant Carlos Hathcock (1942–99) and his observer kept an enemy force pinned down for four days in a paddy field, picking off anyone who tried to flee. Air-dropped

flares illuminated the enemy position at night, while Hathcock changed position to avoid any return fire. Eventually, Hathcock withdrew from the area, but not before calling artillery fire on to the remains of the enemy force.

Hathcock, like many snipers, learned to shoot before joining the military. Hunting to supplement the diet of his poor family, he became adept at both shooting and stalking, skills that made him an ideal candidate for sniper training. The North Vietnamese placed a huge bounty on his head and sent expert snipers to locate and kill him. On one occasion, Hathcock spotted a

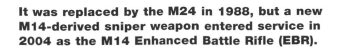

It was replaced by the M24 in 1988, but a new M14-derived sniper weapon entered service in 2004 as the M14 Enhanced Battle Rifle (EBR).

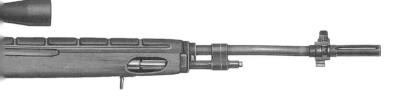

glint of light and fired at what he correctly guessed was an enemy sniper. Upon reaching the target area he discovered that his shot had gone through the scope of the sniper's rifle, which could only have been possible if the weapon had been pointed directly at Hathcock, ready to fire.

Hathcock also achieved what was for many years the longest sniper shot on record, using a converted heavy machine gun. His ability to infiltrate an enemy-held area also set the standard for other snipers. Sent to eliminate a high-ranking North Vietnamese officer, Hathcock set up his shot by crawling 1500m (1640

yards) into a suitable position, which took no less than four days. After killing his target he then had to sneak away again.

Snipers are not psychopaths, and most do not enjoy killing, although the hunt itself can be exhilarating. In many cases, snipers either choose to look at the 'kill' as a job that must be done, or focus on the lives they are saving on their own side. A sniper cannot do anything about the fact that a conflict is going on, and cannot persuade hostiles not to fire on his comrades.

He can, however, protect those comrades: eliminating an enemy soldier may prevent him from killing

Hathcock's .50 cal Sniper HMG

Fitting a bracket to carry a telescopic sight, and converting the weapon to fire a single shot, allowed an M2 machine gun to be used as a sniping weapon. This field expedient was a matter of trial and error, but worked well enough to demonstrate the concept of the large-calibre anti-materiel rifle.

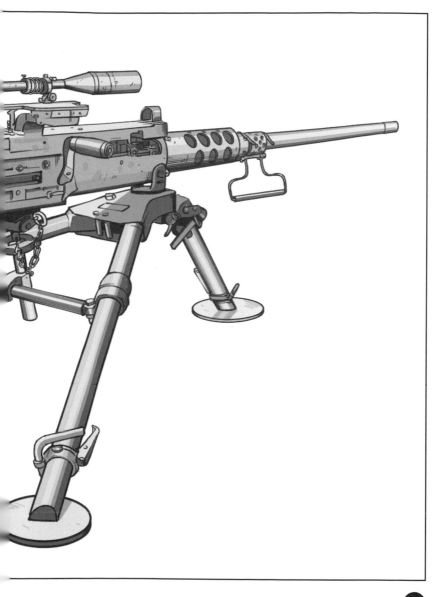

Carlos Hathcock (1942–99)

Nicknamed the 'White Feather' for the feather he wore in his hatband, Carlos Hathcock is perhaps the most influential sniper of all time. His performance during the Vietnam War included almost 100 confirmed kills, including a North Vietnamese general. In order to eliminate this target, Hathcock had to make a long and dangerous penetration of enemy territory, crawling with agonizing slowness into position. Afterwards, he had to extract himself in the same way. His other exploits included shooting an enemy sniper directly down his own telescopic sight and making what was, at the time, the longest-range sniper shot in history, using a converted heavy machine gun. Hathcock and his observer also executed the destruction of an 80-man North Vietnamese patrol. Badly burned while rescuing several Marines from a burning Amtrac, Hathcock became an instructor, teaching both military and police snipers.

any number of the sniper's own side.

This view is shared by many soldiers, who will call on snipers for help and to protect them from enemy marksmen. The poem 'White Feather', written about Carlos Hathcock after his epic infiltration mission, expresses the gratitude of the author for the lives Hathcock saved. The author sees the act of assassinating an enemy officer as one of guardianship, not aggression. It was not the act of killing an enemy that inspired the poem, but fears of what might have happened had that enemy lived: 'There was a battle, never fought, in which I should have died.'

Although this poem was written about the legendary Carlos Hathcock, it sums up a common attitude towards sniping, not least among snipers themselves. The job may be distasteful but it needs doing, and in the long run a sniper will take some lives in order to save many. In the imperfect world of military conflict, this is the best bargain on offer, and a sniper must be at peace with that concept if he is to function effectively.

The Falklands War

British and Argentine snipers played a significant role in the Falklands War (1982). Many Argentine positions included machine-gun teams and snipers, as well as infantry riflemen, working in conjunction to support each another. British counter-sniper techniques included the usual heavy firepower and counter-sniper options, but also MILAN anti-tank guided missiles. This was criticized at the time as wasteful of expensive munitions. Compared to the damage a skilled sniper could inflict, however, missiles were cheap.

Counter-insurgency and Peacekeeping

Sniping and counter-sniping remained an essential part of military operations even when there were no major conflicts ongoing. Counter-insurgency operations in Northern Ireland, Lebanon and other troubled areas demonstrated the need to retain a well-developed sniper doctrine. Similarly, operations such as the NATO intervention in the Balkans during the 1990s were highly complex and required a combination of restraint and precisely applied force; a role to which snipers were ideally suited.

During the Balkan conflict, the city of Sarajevo became notorious for sniper activity, much of it targeted at the civilian population. Snipers positioned in tall buildings could fire down long streets, which became notorious as 'sniper alleys'. Disregarding all the accepted rules of war, many of these snipers chose to

shoot civilians, knowing that others would try to help the victims and so provide additional targets. NATO peacekeeping forces were forced to operate in this environment, often under fire from snipers they could not pinpoint. Massive retaliation with heavy firepower was out of the question in a heavily populated city, so the only answer was counter-sniper fire.

Snipers are often included in a special forces group tasked with a specific mission. The number of personnel that can be deployed is strictly limited; every extra operator increases the chance that the team will be detected before the mission can be accomplished. Thus, the value of any given team member must be evaluated, and a sniper is certainly a high-value asset. Snipers are sometimes deployed to cover the movement of their comrades into the target area and back out again, even if the actual mission does not involve precision shooting.

A large part of counter-insurgency work is observation and intelligence gathering, and snipers have proved to be an excellent asset in this area too. Where a target has to be engaged, a sniper can use a precision shot to ensure that innocents or suspects who have not yet provided justification are not harmed. Their skills enable the security forces to continue to operate within the principles of

Urban Marksman

During the Balkan conflicts of the late Twentieth Century, many self-taught snipers operated in the cities. Some shot from their own apartment windows, using whatever rifles they could obtain. Such snipers often lacked the formal training of a military sniper, and can more properly be considered marksmen than true snipers.

Insurgent Sniper

Roof spaces are an excellent hide for an urban sniper. Removing a few tiles from the roof creates a firing point that can be very hard to spot from ground level. Sandbags can be improvised from any cloth container, and provide both a weapon rest and protection from return fire.

minimum force while responding effectively to a threat.

Iraq and Afghanistan

This has proved especially important when dealing with the counter-insurgency warfare that has characterized recent conflicts. The large-scale military actions of the 1991 Gulf War and Operation Iraqi Freedom in 2003 soon gave way to a much lower level of conflict, as did intervention in Afghanistan (2001 to present). Unable to take on a major military force, the insurgents instead chose to contest control of towns and cities, and to carry out a campaign of ambushes and raids while hiding among the innocent population.

In this environment, major military assets such as artillery, air support and tanks are of limited use most of the time. When they can be deployed against a suitable target, their effectiveness depends greatly upon good reconnaissance and observation of the target. Here, snipers are extremely useful, controlling and calling in powerful support weapons from a concealed position.

While many insurgents in Iraq and Afghanistan were nothing more than highly motivated but unskilled gunmen, snipers proved to be a major threat to the security forces, especially in areas where patrols had to be mounted on predictable routes in urban terrain. As ever, the best

McMillan Tac-50 Rifle

A bolt-action weapon firing 12.7mm (0.5in) ammunition, the Tac-50 has a built-in bipod and has been adopted by the Canadian Army as its primary long-range sniping and anti-materiel weapon. It is also used by the US Navy SEALs.

response to a sniper proved to be effective counter-sniper work, often using teams positioned at key points on the rooftops.

Snipers also supported large-scale operations against insurgent forces in the Afghan countryside. The terrain of Afghanistan offered endless possibilities for ambush, often permitting insurgents to shoot down from concealed positions high above a road. The assault rifles of most infantry, optimized for relatively short-range engagements, were less effective in this environment than was desirable. Sending snipers and designated marksmen (men qualified to a high standard of marksmanship but not fully trained snipers) with convoys increased their ability to shoot back effectively at long range.

For extremely long-range sniping, such as across the valleys of Afghanistan from one hillside to another, large-calibre anti-materiel rifles proved extremely useful. During Operation Anaconda in 2002, a McMillan Tac-50 0.5in (12.7mm) calibre rifle was used to make what were, for a time, the longest shots ever recorded. Three-man sniper teams were fielded around each gun, with each man fully qualified and taking turns as observer, to use the big rifle, provide security or shoot with lesser weapons.

These weapons were used to eliminate enemy snipers, machine-gunner and support-weapon crews at extremely long range, taking some of

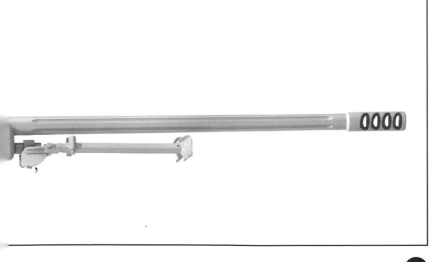

the pressure off friendly forces and allowing pinned-down troops to resume their advance. This exploit required repeated shots of over 1500m (1640 yards), as well as precise shooting when under direct fire. The longest shots made during this operation were by Master Corporal Arron Perry, who hit an enemy artillery observer at 2310m (2526 yards) and Master Corporal Rob Furlong, who broke this record a few days later by eliminating a machine-gunner at 2430m (2559 yards).

Snipers played an important part in countering the threat from improvised explosive devices (IEDs) in both Iraq and Afghanistan. By observing likely places where a device might be deployed, a sniper team could prevent it from being planted. This required great patience, as the snipers waited in a concealed position for something that would, most likely, not happen while they were there. Once in a while they would get lucky and eliminate a bomb team or, occasionally, get the opportunity to take out a senior enemy leader.

Sniper teams were sometimes able to counter attacks with mortars or rocket-propelled grenades. Insurgents who were unaware of the

Operation Anaconda

In March 2002, Coalition forces drove a large force of Taliban fighters from the Shahi Kot Valley in Afghanistan. Insurgents well equipped with machine guns and mortars fiercely resisted the operation. The Coalition forces were supported by joint Canadian–US sniper units, organized as groups of three sharing a McMillan Tac-50 anti-materiel rifle. The snipers used their powerful weapons to eliminate enemy support gunners and snipers while, at times, under heavy fire. Routinely firing at ranges of up to 1500m (1640 yards), the teams broke the world record for the longest confirmed sniper kill twice in a few days. First, Master Corporal Arron Perry hit an enemy artillery observer from 2310m (2526 yards); then, Master Corporal Rob Furlong (b. 1976) eliminated a machine-gunner from 2430m (2559 yards). In the mountainous terrain of Afghanistan, precise long-range support of this nature proved invaluable.

snipers' presence would set up their weapons wherever they had a good line of fire. Since friendly forces were often forced to operate predictably, this in turn made the attacks to some extent predictable. Sniper teams were also able to see through enemy deception in a way that troops in direct contact could not. While the infantrymen who were the target of mortar attacks were primarily concerned with taking cover, a concealed sniper who was not under fire could try to spot and eliminate the observer who was directing the mortar fire.

This also applied to enemy leaders. British sniper Corporal Christopher Reynolds (b. 1984) eliminated the Taliban leader known as Mula during a protracted engagement with Reynolds' parent unit. Having taken up a position atop the roof of a building to support his comrades, Reynolds and his observer noticed a distant figure directing the Taliban forces. Even spotting someone at over 1800m (1968 yards) would be a challenge for most people, but by reading the usual signals (pointing, receiving deference from others and so forth) the sniper team realized that this was the Taliban commander. He was then killed by a shot from 1853m (2026 yards).

Long Shot

In November 2009, Corporal of Horse Craig Harrison (b. 1975) set the current world record for the longest sniper shot. Harrison's troop commander and his comrades were in severe danger, with their vehicle stuck and under heavy fire. Despite a distance of 2475m (2707 yards) between his position and an enemy machine-gun team, Harrison took several ranging shots and then killed both gunners and disabled their weapon, hitting with three out of three shots. The endangered vehicle was subsequently able to escape to safety.

The Sniper Today

The journey from early marksmen to today's expert snipers has been a long one, with occasional setbacks as the sniper's art fell into disregard. However, although the tools have changed and supporting equipment such as laser rangefinders have made some aspects of the task easier, the role and characteristics of the sniper have always been the same. Snipers stalk and wait, take their shot and disappear. They protect friendly troops and disrupt the enemy's capabilities by eliminating leaders, experts and equipment.

The value of snipers is such that it is hard to imagine them ceasing to be a vital part of military forces. Very few people possess the combination of characteristics required to be an effective sniper, but those few have exerted an influence on innumerable conflicts out of all proportion to their numbers.

Modern Sniper

This British Army sniper, seen operating in
Helmand Province in 2010, is wearing a
standard-issue desert camouflage uniform,
with matching Osprey body armour protection.
The latter serves as the replacement for the
previous Combat Body Armour (CBA), and is a
modular system with various ballistic
protection components and configurations. He
is armed with the L115A3, the standard sniper
rifle in service with the British Army.

A sniper is, more than anything else, a keen observer. His telescopic sight allows close inspection of anything that catches his eye, and his ability to spot the tiniest detail that seems to be out of place ensures that little escapes his notice. His response may be a shot, a radio call for artillery or air support, or he may simply continue to observe.

PART TWO:
The Making of a Sniper

2

Snipers are at once part of an infantry force and apart from it. They offer support to and receive support from the main force, but they usually operate separately.

The Sniper's Place in Modern Infantry Doctrine

A sniper cannot function to his best effect when surrounded by a conventional infantry platoon. No matter how experienced or skilled the infantrymen may be, they will inevitably be more noisy and less patient than the sniper, which negates one of his key advantages.

The response of a conventional infantry force to a threat or encounter will tend to differ from that of a sniper. Infantry are trained to be aggressive and to operate as a team, while a sniper will usually prefer to avoid contact and is extremely self-reliant.

Infantrymen usually operate within a fairly well-defined structure and under close command, while snipers are expected to use their initiative within the boundaries of a more general mission.

Attempting to micro-manage a sniper team is a recipe for disaster; instead, it is necessary to allow them freedom of action and to trust the judgement of sniper and observer under unpredictable conditions.

. . . .

A solid weapon rest is a huge advantage when undertaking precision fire. Often a sniper will take advantage of a suitable solid object, which offers some concealment and protection from enemy fire as well as supporting his weapon.

Sniper Cover

Rooftops are favoured sniper positions in urban terrain. They offer a good line of sight, and hostiles are less likely to look up than around them. A weapon rest can be improvised from the sniper's kit if there is nothing suitable nearby.

The Sniper's Mission

The sniper mission is sometimes very different from that of conventional infantry. An infantry unit spends a lot of time on the move, whether on foot patrol or aboard transport, and may have to cover ground quickly in the

course of a mission. Snipers tend to move more slowly and spend a lot of time static. Their engagements tend to be at long range, whereas the average infantry engagement takes place at relatively close quarters.

Infantry combat tends to be a

hand, a sniper will take very few shots and intends to make every one of them count.

These approaches are of course tailored to the sort of combat that is likely to be encountered. At close range, firepower is often more important than precision; snipers are not equipped for a short-range firefight with an enemy squad and will try to leave that sort of engagement to the troops best suited to winning it. Yet there are also similarities between infantry combat and sniping.

In the end, it is rounds that hit the target that decide the outcome of an engagement. The sniper uses stealth and positioning to set up his shot. Infantry soldiers use suppressing fire and manoeuvre to put themselves in a position where the enemy can be successfully engaged, and will use carefully aimed fire when appropriate. The end result is the same; the methods vary by necessity.

Working Together

Thus snipers and conventional infantry complement one another, and often work together. A sniper team deployed to support infantry will usually seek a good position to shoot from and remain there, usually concealed, while the infantry force carries out its mission. A suitable location is as secure as possible, with a good field of fire and, if possible, an elevated position to improve observation. From this

short-range scramble, with a high volume of fire used to suppress hostiles while a rapid dash to a better firing position is made. Engagements take place at a frantic pace, interspersed with sudden lulls of unpredictable length. On the other

vantage point, the sniper team can provide information to the infantry on the ground, and support them if necessary.

The same principles apply whether the troops are operating in the countryside or an urban area. The sniper team remains in position and observes while the infantry carries out its mission. This may involve considerable movement, such as patrolling or conducting house-to-house searches, or may be more static, such as manning a checkpoint.

If contact with hostiles occurs, the sniper team may be able to give warning beforehand, or to report on enemy movements during the firefight. The sniper can, of course, also provide more direct support by shooting enemy combatants. His greatest effect will be achieved if he can identify and eliminate commanders, but by taking out support-weapon crews or enemy riflemen, he can relieve the pressure on the troops in close contact.

The enemy may not suspect the presence of a sniper because their attention will be fixed on the infantry force right in front of them. When a man goes down, it is more common to assume that he has been hit by the troops who are obviously firing at him rather than to look around for a well-concealed sniper who may or may not be present. In this way, the sniper can contribute directly to the

Ambush Sites

'Choke points' on a route offer excellent opportunities for an ambush or sniper attack. Ideally the enemy will be channelled into a narrow area, with limited visibility, and be unable to offer comrades elsewhere in the column effective support due to intervening terrain.

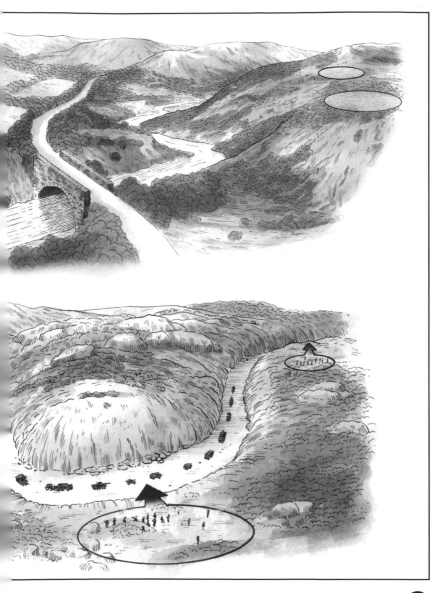

success of the infantry force. He can also call in air support or artillery as necessary, and can provide an overview of the situation to his superiors. His viewpoint is likely to be wider than that of someone in the middle of a firefight.

In such a situation, the sniper acts as support to the infantry force, which will carry out the bulk of the fighting. On other occasions, the sniper may be the only member of the force who fires his weapon. For example, if friendly troops are operating in an area where they are likely to be ambushed or sniped at, the sniper may be the only one who can spot the threat in time or with a clear shot at an enemy marksman.

This kind of complementary operation is extremely effective in keeping control of urban areas. 'Boots on the ground' in the form of patrols and checkpoints are essential in maintaining security and reminding the population that the situation is under control. Infantry can search buildings and arrest suspected insurgents, which requires a hands-on presence. A sniper might make the enemy fearful of operating in a given area, but he cannot gain long-term control in the same way.

What a sniper can do is gather intelligence on suspected insurgents and provide additional security for the troops on the ground. A patrol moving through a town is an inviting target for an enemy sniper, who will

almost certainly choose a position in which he is hard to spot and counter. A supporting sniper can watch such positions from his own vantage point and, ideally, eliminate the hostile before he can take his shot.

Sniper Team Support

Personnel carrying out tasks such as mine clearance or engineering works are vulnerable to enemy attack. A watchful sniper may be able to give warning, and will certainly be well positioned for effective return fire.

Designated Marksmen

Snipers can also be embedded in an infantry force to provide long-range precision fire when needed. This is something of a waste of their capabilities, and so this role is normally given to designated marksmen. These are highly skilled sharpshooters who share some skills and training with snipers, but generally lack the specialized stealth and observation skills of the true sniper.

Designated marksmen are specialist infantry soldiers, in much the same way that others may specialize as grenadiers, communications experts or support gunners. They are part of a unit and have the same duties as anyone else, but are armed with a highly accurate long-range rifle. Whereas some snipers prefer bolt-action weapons to semi-automatic rifles, a designated marksman will normally use a semi-automatic weapon and can lay down rapid fire if necessary.

It is common for designated marksman weapons to use a relatively heavy-calibre round compared to the assault rifles of the infantry. A marksman rifle is typically chambered for 7.62mm x 51mm (0.3in calibre) rather than the more common 5.56mm x 45mm (0.219in) cartridge used by assault rifles. This allows accurate shooting at longer ranges, and marksman rifles will usually have a telescopic sight.

The Soviet Army, and the armed forces of the post-Soviet states, have long made use of marksmen within an infantry force. Each platoon contains a man armed with an SVD rifle rather than the AK assault rifles that arm most infantrymen. The marksman's task is to pick off high-value targets such as radio operators and officers while his comrades engage the enemy force as a whole.

Other armed forces were slower to adopt the designated marksman concept, but today it is an important part of infantry capability. Long-range shooting has proven to be important in Afghanistan, where engagement ranges tend to be longer than on average. Weapons and training of most forces are optimized for fairly close-range combat, often in urban terrain, and can be less effective when dealing with an ambush in the mountains of Afghanistan. Picking off insurgents armed with rocket-propelled grenades is particularly important.

Security Support Roles

Snipers are sometimes used to provide security for a force using a predictable route. A sniper team can be dropped off as the patrol or convoy heads out towards its objective and will take up a suitable observation point. If hostiles attempt to plant explosives or set up an ambush, the sniper team can warn of the danger or engage if conditions seem favourable. The sniper team may be picked up as the convoy proceeds back to base, or may be recovered by other forces.

Snipers are also integrated into standing security operations. They may be deployed inside a base or defended area, or posted at a suitable point outside. In most cases this means long, boring hours of watching a road or a stretch of desert in case something happens. Sometimes the sniper's skills will

Squad Marksman Weapons

Squad marksmen are not snipers, but they are very good shots. Armed with an accurate rifle, a marksman extends the squad's effective engagement range enormously. Below are a selection of the more popular marksman rifles of the last 30 years.

Dragunov SVD

M14 EBR

M40A1

enable him to identify suspicious patterns of behaviour, such as disguised enemy personnel reconnoitring the base. More often, the sniper is there 'just in case', but nothing happens.

Sometimes, however, the sniper is able to intervene to assist his comrades. Gate guards, checkpoints and other standing security arrangements cannot be fielded in the sort of strength that would be deployed to meet an expected attack, which means that the security detail may have to deal with an incident using whatever scanty

Overwatch Security

A sniper who has to support a heavy rifle for an extended period, and wishes to move little in order to reduce the chances of being spotted, will assume an easy-to-maintain posture. This may or may not actually be comfortable.

resources are available, at least until assistance arrives. The support of a sniper can make all the difference in such a situation.

Attacks on security checkpoints, base gate guards and the like can take many forms. Among them are personal and vehicle-borne suicide bombs, larger-scale attacks, mortaring and of course sniping. While a sniper situated near the target cannot usually do much about a distant mortar crew, he may be able to eliminate an observer who is reporting the fall of shot. Other attacks can be broken up or stopped entirely by a well-placed shot. This is especially important in the case of suicide bombs; the earlier the bomb carrier is disabled, the less likely that friendly personnel will be harmed.

Defensive and Offensive Roles

When a sniper team is deployed in support of other friendly troops, its role is as at least as much defensive as offensive. The team may engage suitable targets but, as a rule, it is present to assist the infantry and to protect them from threats such as enemy machine-gunners and snipers. Generally speaking, protection and assistance of the infantry force is the sniper team's primary role and any opportunity to engage the enemy must be weighed against the loss of this protection.

A team that is deployed away from friendly troops generally has a purely offensive mission. The goal may still be to keep an area clear of hostiles, i.e. to defend it, but the sniper team is not concerned with supporting friendlies in this case. Instead, the sniper takes the fight to the enemy. Missions of this sort include a deliberate assassination of key

Checkpoint Support

A sniper equipped with a powerful rifle can halt a vehicle well short of the checkpoint with a shot to the engine block. This can be critical if the vehicle is rigged with explosives, or when there is doubt as to whether the occupants are hostile or merely frightened.

enemy figures, the removal of an enemy strongpoint or observation post, and of course counter-sniper work.

When a sniper team can operate independently in this manner, the sniper is free to decide how best to accomplish his mission. Reaching the

example, if an enemy observation post is located in difficult terrain, it may not be possible to remove it without deploying a fairly large force, which may result in casualties if the observation post is well defended. On the other hand, a sniper team may be able to eliminate the observers from a distance or sneak past the defenders to make a close-range shot.

While designated marksmen are part of an infantry unit, snipers are normally grouped into a sniper platoon. The platoon will rarely, if ever, deploy together but acts as an administrative unit. Sniper teams can then be assigned as needed, either independently or in support of other forces.

Grouping snipers in this manner makes training and equipment maintenance simpler, and ensures that the sniper force is commanded by someone who understands their unique nature. Snipers have a different mindset to the typical infantryman and may not mix well; friction between team-oriented infantrymen and the 'loner' sniper is not desirable. There are other philosophies about how to organize or command snipers, but the overall rule is that they are flexible assets that must be given sufficient autonomy to make best use of their skills, so their place in the structure of a force must not be so rigidly defined as to limit their effectiveness.

area and stalking the target may take a long time, but ultimately the sniper may be able to accomplish tasks that an infantry force could not. For

3

Although the term 'sniper' is used in both police and military contexts, the role and methods of operation of police snipers are somewhat different to their military counterparts. While police marksmen prefer not to be observed by the target, they will rarely use the same sort of stealth and concealment skills as a military sniper. More commonly, they will move into position during an ongoing situation, sometimes quite openly. Indeed, letting a criminal know that police snipers have been deployed may be used as a psychological gambit to induce surrender. ① to gain an advantage

② Police snipers may be deployed as a security measure, for example along the intended route of a VIP, or they may be brought in as a response to a hostage taking. They will normally deploy in an elevated position with a good line of sight to the intended target, and rarely need to be very concerned with concealment or the prospect of counter-sniping. Any situation where criminals are willing and able to shoot it out with police snipers has become very serious indeed.

• •

The role of the police sniper is different from that of their military counterpart, requiring close co-operation with other law enforcement officers and strict rules of engagement.

Snipers in Law Enforcement

Suspect Gone to Ground

Someone who has gone to ground after committing a crime that requires an armed response is likely to realize that snipers will be deployed. A suspect may well be able to figure out where the marksmen are most likely to set up, but he will not usually be able to do anything about it. Most commonly, the fugitive will be contained or may have taken refuge voluntarily. His location will be known and the sniper can deploy accordingly. This further limits the suspect's options; once snipers have covered the exits from his position, then escape is very unlikely.

Police marksmen are empowered to use lethal force under fairly specific conditions. Whereas a military sniper

SG-1 Sniper Rifle

Based on Germany's highly influential G3 infantry rifle, the SG-1 differs mainly in the use of a heavy barrel and an improved trigger group. Equipped with a bipod as standard, the SG-1 found favour with several police agencies and some special forces formations.

in a declared war is free to engage any enemy combatant regardless of whether he poses an imminent threat to friendly forces or innocents, the police sniper is limited by rules of engagement similar to those binding troops deployed as peacekeepers.

Last Resort

A police sniper will normally be permitted to open fire in order to protect himself, fellow officers or any other person from imminent harm; and in some cases to prevent the escape of a dangerous individual who has demonstrated the ability and willingness to use violence. Whereas for the military sniper, taking the shot is the focus of the mission, police snipers are part of a

Justified shot

Enfield Enforcer

Developed for law enforcement use from the Mk 4 Lee-Enfield infantry rifle, the Enforcer had the same origins as the military L42A1. It used different furniture, notably the stock and grip.

team that will try to apprehend criminals without violence, if possible. Thus shooting is the absolute last resort.

In general, police snipers operate in an urban environment and shoot from within 100m (109 yards) of the target. This presents some unique problems. As already noted, a shot must be justified at the moment it is fired, rather than being directed at a declared enemy who is a legitimate target simply by being involved in the conflict.

In addition, the sniper must be careful about where his bullet might go if it passes through the target. He may have to shoot through a window or past a hostage.

In the case where hostages have been taken, a police sniper will have to differentiate between hostages and hostage-takers, which can be

dragging

difficult even if the suspects are not deliberately trying to deceive observers. He may have to watch the suspects for a long time, which is as tiring as it is tedious, and then make an instant decision whether or not to shoot, and who to target.

Intelligence Gathering

Police snipers spend far more time watching a suspect than they do shooting, and are often employed as intelligence-gathering assets much like their military counterparts. From an elevated position, a sniper can often see more than officers on the ground, and can also act as early warning that circumstances are about to change. This might happen because of something the suspects are doing or due to the arrival of others from outside the police cordon.

Marksmen will generally shoot in

an encircle line

SWAT Team

A cornered suspect may be able to hide from a police sniper but will not be able to flee without being engaged. With the situation as firmly under control as possible, an entry team takes over the task of bringing the situation to a close.

response to something the target is doing, which usually means they have a very brief moment to assess the situation, aim and shoot before the opportunity is lost. Hesitation or a missed shot can mean an officer or a hostage killed, and shots must often be extremely precise. For example, if the suspect has a weapon pointed at a hostage, the sniper must ensure he is killed instantly and unable to pull the trigger. This requires a shot into a small area of the head or spinal column, which is difficult to achieve at any range.

Law-enforcement marksmen are sometimes called upon to fire on

Hostage Rescue

The deployment of sniper teams can be used to contain a situation while a solution is negotiated. If the suspects begin to harm their hostages, the sniper team must act instantly to save lives. The leap from monotonous observation to decisive action is a hard one to make.

non-human targets. These might include dangerous animals as well as a range of inanimate objects. Commonly, marksmen are called upon to disable vehicles by targeting the engine with a powerful rifle. The US Coast Guard (an organization with both military and law-enforcement characteristics) has used this technique to halt suspected drugs-carrying boats by shooting out the engines. This may be justified in circumstances where firing on the occupants of the vehicle would not be, although the sniper has to be very careful when engaging a moving vehicle with a large-calibre rifle.

Police Sniper

Police snipers generally shoot from a relatively short distance, and rarely need to be concerned with stealth or concealment like their military counterparts. However, they have to operate in a cluttered urban environment, and must always be mindful of nearby innocents.

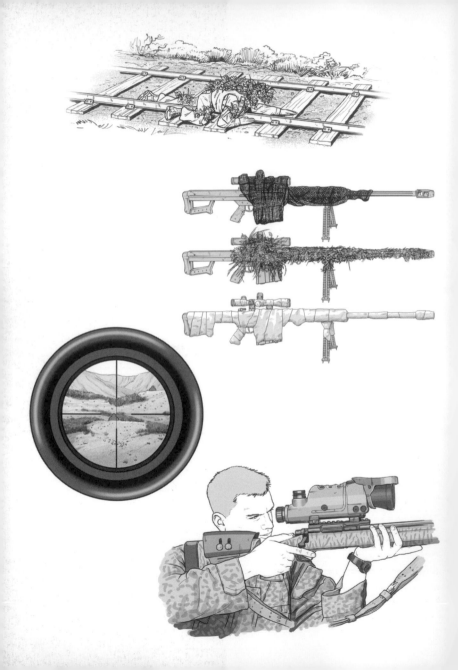

Most modern armed forces maintain a sniper training programme organized along broadly similar lines. Potential snipers come to the school after proving themselves in another role, usually as infantrymen. The competition just to be selected to attend the school is fierce, and most candidates do not pass the course.

Those who do graduate become snipers, while those who do not return to their units with a wealth of new knowledge and experience, which raises the overall standard of the force. Therefore, the sniper school is more than a way of creating snipers; it is also a centre of excellence for infantry skills.

The Right Qualities

In order to be considered for training as a sniper, a soldier has to demonstrate the right qualities. These include an ability to shoot well, of course, but also habitual observation and the ability to make solid deductions from those observations. He must also be intelligent, self-reliant and confident. Often it is not enough to observe and report and then ask for orders; the sniper must decide how best to act on his

. . . .

A successful sniper must be a master of concealment, and must be able to visualise what an enemy searching for him is likely to see.

4

Sniper selection is more than just a matter of finding the best marksmen. He must also possess the correct mindset, without which he cannot function.

Sniper Selection and Training

observations and then formulate a plan accordingly. He risks his life and the lives of others on this decision, so it is vital for potential snipers to possess sufficient conviction to make the best decision possible under the circumstances and act on it with determination.

Using a Scope

Placing the crosshairs on a target does not guarantee a hit. Even if it is perfectly adjusted for range, a telescopic sight only indicates where the weapon is aimed. Wind, humidity and air temperature may all require an adjustment to the aim point.

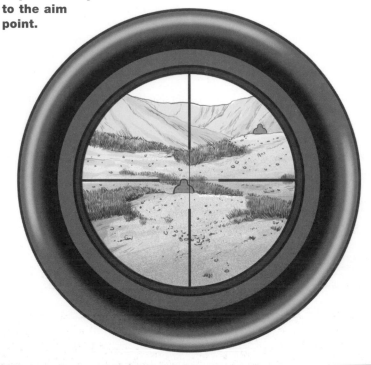

Those lacking the necessary skills and psychological attributes are usually screened out at the application stage. If, somehow, they manage to reach the sniper school, they rarely last long – and nor do those who think they can coast through the course. Everyone who attends sniper school is already highly skilled, but the course is designed to build on that high starting level – being a well-above-average soldier is an entry requirement, not a sure ticket to success.

The sniper course is intense, and not merely because time is limited. Sniping is a combination of high stress and inactivity, which most people do not handle well. Action is an excellent way of coping with stress, but a sniper who cannot wait for the patrol to pass or the right moment to shoot will at best be unsuccessful.

Candidates are given increasingly challenging tasks to perform, and the reward for success at a difficult exercise is to be sent to do the near impossible.

The learning curve is steep even for the best of applicants, and this too is a test of the candidate's resolve. A successful sniper quietly does his job then moves on; his work is not often witnessed or appreciated by others. Someone who feels aggrieved that he is not receiving due recognition is not temperamentally suited to sniping.

Technical and Academic Skills

A significant part of sniper training is technical and academic. Snipers must be able to operate communications equipment as well as maintaining their own rifles and accessories. They must also understand the effects of wind, humidity, temperature and gravity on a shot. Specialist knowledge includes the effects of shooting through different types of material, notably glass, and the ability to

'No Psychos'

Sniper selection is designed to weed out 'psychos' who just want to kill people, and any that slip through are unlikely to last through the highly disciplined training period. A psychotic killing rampage is the very antithesis of the sniper's mindset: his is a clinical job performed with calm and precision.

Bunker-Type OP

A bunker-type OP consists of a large excavated space, enough to house several men, lined with sandbags or plastic sheeting to strengthen the structure and to keep out water. It is covered with a substantial wooded roof.

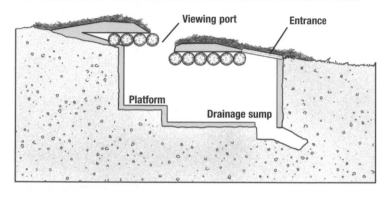

Viewing port   Entrance

Platform

Drainage sump

judge whether a given piece of cover will stop a bullet or not, which may affect the decision whether or not to shoot at that moment.

Snipers must also be thoroughly versed in procedures ranging from calling in air support to reporting an observation, possibly to friendly forces that are not the sniper's parent force. An understanding of how friendly troops operate helps the sniper predict what friendlies in the area might do, and judge how best to support them. At the same time, knowing common search techniques is essential to defeating them. On top of all this, it is necessary for the sniper to be able to recognize key personnel by their behaviour, insignia or general characteristics, and to identify enemy equipment in order to make a clear and correct report of sightings.

Field Craft

Snipers are also taught field craft, i.e. the ability to operate in the field while remaining both unobserved and aware of their surroundings. This includes obvious skills such as the ability to construct a hide but also an understanding of the interplay of light and shadow, and the ability of the human eye to

Using Light and Shadow

Shadows can conceal a sniper or any object, but they can also give away an otherwise well concealed position. A sniper is always mindful of what is behind him, and how the light falls. Standing out as a clear shadow is just as bad as being brightly lit.

distinguish patterns and shapes. Once the sniper understands how the brain recognizes shapes and builds a picture accordingly, he can learn how to fool it or at the very least how to avoid presenting an

Blending In

Even in the open, it is possible to conceal the human shape by blending in with the surroundings. Straight lines such as a fence or railway line can confuse the eye into missing an outstretched limb lying along them.

observer with obvious clues that will attract his attention.

This leads into the principles of camouflage and concealment, and how to break up the telltale shapes of the head and rifle. Snipers learn to use a ghillie suit or other camouflage to turn themselves into a shapeless mass the same colour as the background. They learn too how to select pieces of vegetation to add to their camouflage so that it blends in with the surroundings.

Conversely, sniper candidates learn to estimate range and to distinguish objects even when they are camouflaged. They also participate in several 'stalks' against observers who are trying to spot them. These exercises do more than build a habit of careful observation; they help snipers learn how not to give away their position or presence.

Accurate Shooting

Of course, snipers also train to shoot accurately at various distances on the firing range and in various field conditions. A sniper must be able to instinctively 'lead' a moving target the right amount, and must account for the fact that a bullet may take a couple of seconds to reach the target when shooting at

Breaking up the Shape of a Rifle

Rifles have many straight lines, which are rare in nature. Netting, sacking or foliage can be used to soften the outline of the rifle, making it more likely to be mistaken for a branch or similar natural object.

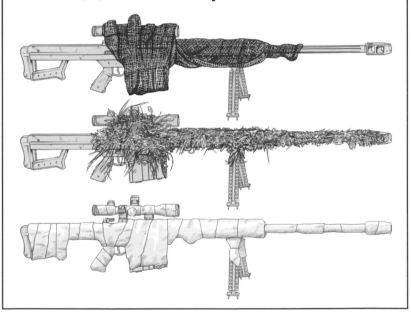

long distances. This requires a great deal of practice, but before these skills can be of any use, the sniper must first learn to shoot precisely from a variety of positions.

Accurate aimed shooting is, to a great extent, dependent upon keeping the weapon absolutely steady while aiming and firing. Snipers learn to support the weapon in a variety of ways, using their body and solid objects or the ground. Breathing must be controlled, as expansion and contraction of the

Squeezing the Trigger

Sniping weapons have a very light trigger break, but even so a sloppy pull can drag the weapon off target at the instant of firing. A sniper squeezes the trigger rather than pulling on it. He uses the pad of his finger and strives to move nothing other than his trigger finger, even holding his breath to avoid disturbing the shot.

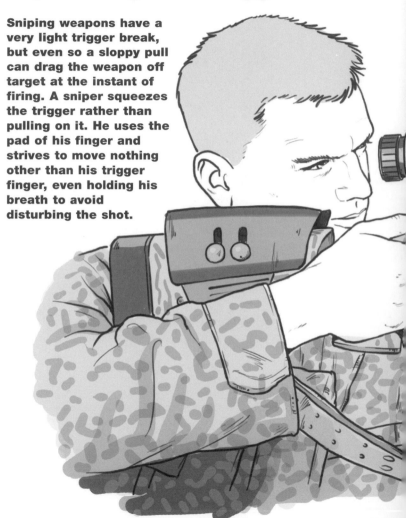

chest will cause a rifle to wander off target, so snipers learn to hold half a breath while setting up a shot. However, as a result, they will then have only a very limited period before the blood starts to become deoxygenated and the rifle starts to tremble. A skilled sniper knows instinctively how long his period of maximum accuracy will last – it is only a few seconds – and he will either take the shot in this period or abort and set up another.

The mechanics of marksmanship are endlessly practised, including seemingly trivial skills such as how to squeeze the trigger correctly. This is, in fact, actually of immense importance: the trigger must be activated without moving the rest of the weapon in any way, and a sniper must learn not to 'snatch' a shot in this manner. Like much of a sniper's training, this is as much about learning to put aside the adrenaline and stress that are part of taking a shot, and learning how always to make a mechanically perfect shot.

Put to the Test

Among the challenges facing a prospective sniper is an observation test in which various items of military equipment are concealed and must be spotted by the sniper within a set time frame. The sniper must also complete a stalking exercise in which he has to infiltrate close to a position occupied by a team of qualified

Observation Test

snipers who are trying to spot him. He must succeed in taking two shots with blank cartridges without being detected. This last challenge is a nightmare scenario – the sniper must operate closer to the 'enemy' than he normally would, and that enemy

Snipers' powers of observation are routinely tested to a point far beyond what a normal person could achieve. This is not only to hone the sniper's own skills; he must also understand what an observer is likely to see and what they will almost certainly miss.

is a team of experts who know everything he does. They also know that he is present, and what he is trying to do. Anyone who can achieve success under these conditions deserves his qualification and graduates from the course as a sniper. Of course, this is not the end of his education. Most militaries pair up a newly qualified sniper with an experienced mentor who will continue his training on the job until, one day, he is experienced enough to take on a rookie of his own.

Much of a sniper's equipment is fairly mundane, and little different from that of the conventional infantry. He has much the same needs for water, food and basic tools such as a knife or eating implements. It is likely that he will not wear body armour or a helmet much of the time, as these are more likely to be an encumbrance than an asset. Snipers do not intend to get involved in short-range firefights, and defend themselves better by concealment and breaking contact than by armour.

A sniper will usually have an optical device separate from his rifle, for use in spotting targets or conducting general reconnaissance. This might be a simple pair of binoculars, or a sophisticated instrument such as the 20x scope used by the US Army. Snipers sometimes make use of thermal imaging or low-light equipment to assist in finding targets at night, and may use a periscope to see over hard cover without exposing themselves.

Camouflage

Camouflage equipment is also extremely important. The sniper's

....

Snipers can use a wide range of equipment, both to improve accuracy and enhance concealment. Often, the best camouflage materials can be obtained from a nearby bush.

5

Snipers make use of a variety of equipment. Some is precision manufactured, while many other items are improvised in the field.

Sniper Equipment

Helmet Camouflage

By attaching vegetation to his helmet, the sniper can break up his head silhouette. The key to such camouflage is to keep it fresh and to ensure that it matches precisely the terrain through which you are moving.

fatigues and head covering (usually a hat or beret) are either in a camouflage pattern or at least of an unobtrusive colour that will not stand out. Camouflage can be added on an ad-hoc basis, such as small branches around the head and

shoulders to break up the outline, or other vegetation loosely attached to the sniper's clothing.

Ghillie Suit

Many snipers use ghillie suits, although they are not always

Face Camouflage

The 'blotch' pattern of face camouflage here is ideally suited to temperate deciduous (leaf-shedding) areas, desert and barren, snowy landscapes. Don't overdo the dark areas – too much dark camouflage can make you stand out even more.

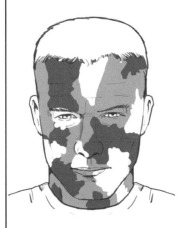 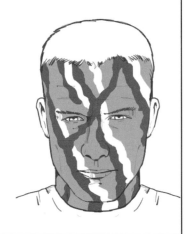

appropriate. A ghillie suit is loose fitting with scraps of cloth attached that move and fall naturally, much like the vegetation they mimic. Obviously, as a ghillie suit must be tailored to the environment in which it is to be used, it is of a common base colour with strips of varying hue attached. A mass of a single colour, no matter how little it looks like a person, will attract the eye, so a ghillie suit must have a natural colour variation to blend in with local vegetation, natural features and shadows. The ghillie suit can be adapted by adding local vegetation as the sniper moves along, which allows him to match his suit exactly to conditions in the area he is passing through. While this takes time and slows down movement, it is an extremely effective method of camouflage.

Rifle Concealment

No matter how well camouflaged the user may be, his rifle must also be

Modern Ghillie Suit

A ghillie suit turns the sniper into a shapeless mass which is not readily identifiable as a human being. This is pointless if the rifle is not camouflaged, however; a bush armed with a rifle will soon be recognised for what it really is.

concealed. Straight lines draw the eye, as they do not occur in nature. The sniper must therefore camouflage his weapon and break up its outlines. Most obviously, the rifle (as well as other metal objects the sniper carries) must not be shiny and reflective, and will often be painted in camouflage colours suited to the environment. Scrim net or other loose cloth is often used to further hide the outline of the weapon.

Thermal Imaging

One threat faced by modern snipers is the use of thermal-imaging technology. No amount of conventional camouflage can conceal the 'glow' that emanates from a warm human body and might make a sniper easy to spot – but there are countermeasures. Specialist materials are available that vastly reduce the amount of heat the sniper radiates, making him much harder to spot. It is also possible to reduce a thermal signature by simple methods such as good positioning; a thermal imager cannot see through a solid object.

Weapons

A sniper may carry a back-up weapon. Usually, this is a sidearm of

Sidearms

Most snipers carry a handgun, usually a high-capacity 9mm (0.354in), for emergency use. A bolt-action rifle lacks the firepower for a close-range engagement, but a handgun is not much better. If a sniper is down to his sidearm, something has gone horribly wrong.

SIG Sauer P228

Beretta M9

some kind – normally a handgun – but there are times when a sniper will also take with him a carbine or sub-machine gun. His observer will normally be armed with an assault rifle. Some forces prefer to give the observer a sniper rifle of his own, working on the assumption that this effectively doubles the number of snipers available. Most, however, accept that the sniper and observer have different roles, and that if a close contact occurs, then a partner with an automatic weapon, and

Types of Sniper Rifle

There are arguments in favour of both semi-automatic and bolt-action sniping weapons. A bolt-action rifle is arguably a better choice for extreme long-range sniping, while a semi-automatic weapon allows rapid engagement of several targets.

Accuracy International
L96A1

possibly a grenade launcher, will be of more assistance than a second sniper.

Bolt Action vs Semi-Automatic

The sniper himself will be armed with a specialist rifle. Traditionally, this was a bolt-action rifle but there is currently a move towards semi-automatic weapons. There are arguments both ways, and differing schools of thought about which is best. One reason for the traditional use of bolt-action rifles is no longer

M39 Marksman Rifle

To some extent, weapons can be tailored to the mission but it is the sniper, not the weapon, that is most important. A sniper will carry the rifle he can perform best with, regardless of what works for someone else.

really valid today. Semi-automatic weapons were usually mass-produced for infantry use and, while accurate enough for the average soldier, were not ideal for sniping.

Excellent examples could be made from scratch, but it was more common to seek an accurate rifle among those already available, which usually meant high-quality civilian hunting rifles. Consequently, most sniping weapons were based on bolt-action weapons sourced outside the normal military procurement system and adopted into service in small numbers.

Today, this is not necessary. It is possible to manufacture semi-automatic weapons that are inherently as accurate as any bolt-action rifle, except in one way. A semi-automatic weapon has moving parts at work as soon as a round is fired. While it is still proceeding down the barrel, the action is working. While the movement of a gas piston or bolt is small, it does represent a change in the weapon's weight distribution and some internal momentum. This may affect the aim point marginally as the weapon is fired, though the effect is likely to be minimal and only significant when shooting at extremely long ranges.

A bolt-action rifle, on the other hand, fires from a bolt that stays closed and locked until the firer chooses to work the action. There is no movement to disturb the aim point, and more importantly no automatic ejection of spent cartridges. A semi-automatic weapon spits a shiny metal cartridge case out when it fires. This may glint in sunlight or be spotted as it flies out of an otherwise innocuous bush.

Spent cases may need to be collected if the sniper wishes to keep his location secret, i.e. to prevent patrols from discovering where the shot came from and

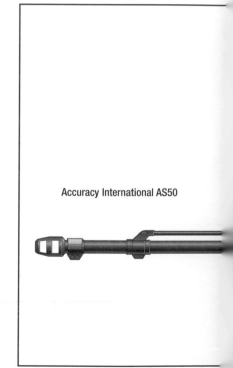

Accuracy International AS50

setting an ambush. Snipers prefer not to use the same location repeatedly but sometimes have no choice. Cartridge cases may also give the enemy at least an idea of which force sent the sniper, which may not be desirable even if they can discern nothing more. Thus, spent cases must be collected, and this requires the sniper to move – at least a little – possibly at a time when he would prefer to remain still and hidden. On the other hand, a bolt-action weapon can be opened slowly and quietly and the spent case removed by hand. However, it is necessary to take a hand off the weapon in order to work the action, and this will disturb the aim point as well as slowing down the rate of fire. There are times when rapid shots are desirable, such as when engaging multiple hostiles or following up a missed shot on a moving target. A semi-automatic weapon also allows the sniper to engage fairly effectively

AS50 Rifle

A British-made 12.7mm (0.5in) calibre rifle, the AS50 has a free-floating barrel; a feature found more commonly on much lighter weapons and rarely in an anti-materiel rifle. At the rear of the stock is a folding handgrip/monopod, and the weapon is supported at the fore end of the receiver by a folding bipod.

Free-floating Barrel

A free-floating barrel is not attached to the forearm of the rifle, but is supported only where it meets the receiver. This prevents the barrel from being pulled out of alignment by the forearm, but it does mean that the weapon is more fragile than a conventional design.

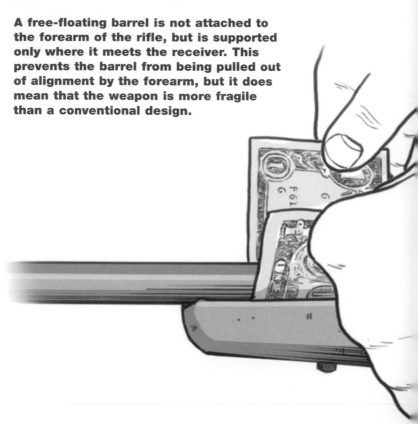

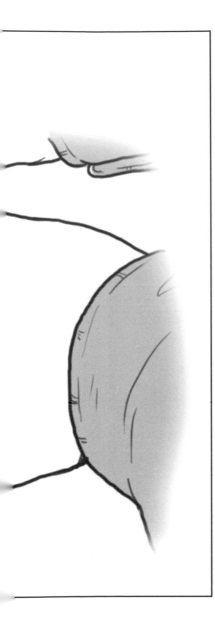

in normal infantry combat, which can be important if the team is discovered or is part of an infantry force that is engaged while the sniper is in close proximity.

Rifle Characteristics

While there are differing schools of thought about what kind of action a rifle should have, rifle characteristics tend to be quite similar. A rifle must, of course, be extremely accurate and capable of retaining this level of accuracy even when operating in the field for an extended period. Components must be of the highest quality and machined very precisely. The weapon must be assembled by a craftsman and maintained by an expert. However, there are numerous ways to achieve this.

Free-floating Barrels and Muzzle Droop

Many sniping weapons have a 'free-floating' barrel. This means that the barrel is attached only to the receiver, and has no contact with the forearm of the weapon. This eliminates the possibility that tension in the forearm or at the points of attachment may cause small distortions in the barrel. The forearm itself extends under the barrel and can be used to support the weapon as a whole rather than pushing the barrel up.

A free-floating barrel relies on the unsupported strength of the barrel to retain its shape. This is not a serious problem in a fairly short weapon but a very heavy or long gun can suffer from muzzle droop as the weight of the barrel causes it to sag slightly as distance from the support increases. A hot rifle will droop more than a cool one, and one that is supported along the barrel or near the tip will suffer less.

M24 Sniper Kit

At the heart of a sniper's equipment is the rifle, but its effectiveness would be greatly diminished without the rest of the kit. Keeping the weapon clean and properly maintained can be difficult in the field, but it is essential for a first-shot hit.

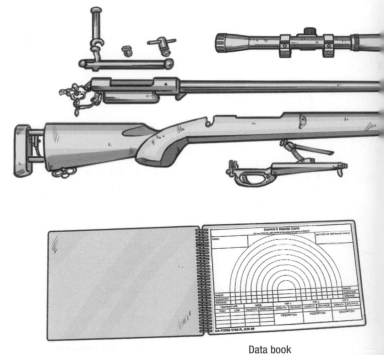

Data book

Muzzle droop is not a serious problem in the typical rifle used for sniping, so a free-floating barrel is generally considered a good trade-off. However, some very fine weapons support the barrel at one or more additional points, so the argument is not necessarily clear-cut.

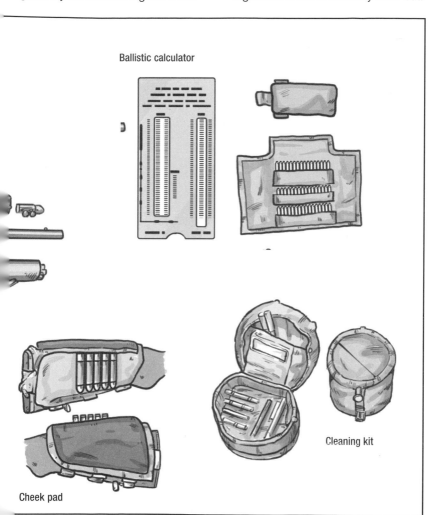

Ballistic calculator

Cleaning kit

Cheek pad

Recoil

The barrel of a sniping weapon tends to be somewhat heavier than that of a typical combat rifle, which helps dissipate heat and absorb recoil. The latter is a serious problem when shooting at long distance, as the weapon's movement may disturb the aim point before the round has left the muzzle. Recoil is minimized in a variety of ways, some of which are inherent in the weapon's construction.

Any rifle is a fairly weighty object, which helps to reduce the amount of movement caused by firing a round.

Physics dictates that if a small, relatively light bullet is fired from a weapon at high velocity in one direction, then the weapon itself will move the other way. Being much heavier, it will move far slower, of course.

As far as possible, a rifle is designed to cause the recoil of firing to push directly backwards, as this does not change the aim point. A well-balanced weapon will move backwards but the distribution of weight as well as the way the weapon is held may cause the

Walther WA2000 Rifle

Using an unusual bullpup configuration, the WA2000 is one of the most accurate rifles ever produced, but it is also extremely expensive. It is intended for the law-enforcement market rather than military sniping, for which it is insufficiently robust.

muzzle to rise as well. This is partly counteracted by the sniper's hold on his weapon, but good design can make a given weapon much easier to control.

Robustness vs Precision

Another trade-off in weapon design is the need to balance robustness with precision. A Formula One racing car may be able to outpace any other four-wheeled vehicle on a suitable track, but it may not perform so well in a street race. The same applies to rifles: the very finest are also usually

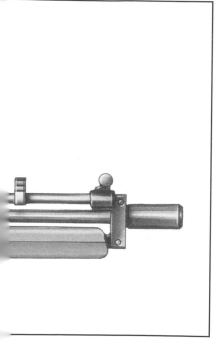

the most fragile. Some rifles, such as the WA2000, are incredibly accurate but not well suited to the rigours of field service. The Galil Sniper Rifle, on the other hand, is far less accurate under perfect conditions but will function well even when treated to appaling abuse.

If the weapon is to be carried in a case, protected by padding, and only brought out when it is time to shoot, it does not matter if it is not very resistant to being knocked out of alignment. On the other hand, a rifle that is going to be carried over rough terrain, and perhaps bounced around in vehicles, needs to be capable of retaining its accuracy under these conditions. There are different approaches to this trade-off, though most sniping weapons are a good balance of toughness and excellence.

Capabilities vs Portability

A further balance must be struck between the weapon's capabilities and portability. A long barrel gives a higher muzzle velocity, which improves accuracy and range, but adds to the weight of the weapon. It also causes other problems. A long rifle can be an encumbrance when operating with other troops in vehicles or urban terrain, and can be awkward when moving through close terrain such as forest undergrowth. A long barrel is also harder to conceal and may be prone to the effects of

Anti-Materiel Rifles

Anti-materiel rifles are sometimes used with specialist ammunition, such as high-explosive or incendiary rounds. Even standard ball ammunition is enormously destructive when fired from such a powerful weapon, and is sufficient for many tasks.

PGM Hecate II

Gepard M6

One of the most potent anti-materiel weapons in the world today, the 14.5mm (0.57in) Gepard M6 uses a five-round magazine and has a bullpup configuration to keep its length manageable. It is 450mm (17.5in) shorter than its predecessor, which was chambered for the same ammunition but used a conventional layout.

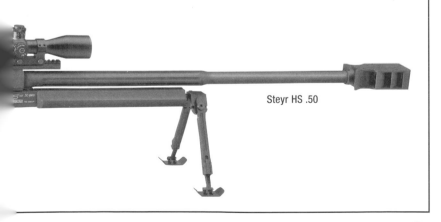

Steyr HS .50

wind; even a slight breeze can push the weapon muzzle around a little – and a little is all it takes to ruin a critical shot. One way to increase barrel length without altering the overall size of the weapon is to use a

Galil Sniper Rifle

Based on the Galil assault rifle, itself derived from the AK series of assault rifles, the Galil Sniper is perhaps better suited to the marksman role than true sniping, but it is a very robust and reliable weapon, and works well within its limits.

Mass and Velocity

Bullets cause damage and injury by delivering kinetic energy (energy of movement) into the target. The amount of energy possessed by a bullet depends on its mass and, more importantly, its velocity. However, it is the ability of the round to deliver this energy that matters. A soft bullet that deforms in the wound will shed its energy quickly, delivering more of it to the target in a shorter time.

'bullpup' configuration. A bullpup weapon is fed from a magazine located behind the trigger group rather than in front of it as in a traditional rifle.

This configuration can effectively shorten a rifle or allow a longer barrel on one that is the same size as a conventional weapon, by moving the feed mechanism into the stock. However, this does alter the balance of the weapon, which may not be desirable. There are few bullpup sniping weapons, but those that do exist include some of the finest rifles in the world.

Calibre

Calibre is another important question. There are few small-calibre sniping weapons; most are in the 7.62mm (0.3in) range, as a heavy round gives a longer range and greater stability in flight, as well as superior penetration and lethality. However, a heavy round also requires a heavier rifle. To this must be added the weight of a heavy barrel, sighting equipment, a bipod or monopod, and any other accessories the weapon may be fitted with. So, even a fairly standard sniper rifle is a fairly large and heavy piece of kit.

A rifle chambered for a very large round such as the 12.7mm (0.5in) BMG (Browning Machine Gun) round, for example the M82 Barrett, will of necessity be very large and heavy. The receiver needs to be much greater in diameter to contain the

12.7mm x 99mm round (as opposed to a 7.62mm x 51mm for a typical weapon). This almost doubling of ammunition and barrel diameter translates to a lot more metal in the weapon. The sniper has to carry this.

A huge round like this also requires a very robust weapon, which again means more material used to construct it, and it may need a special recoil system to absorb the fearsome energy of the powerful round. The M82 Barrett uses an artillery-style recoil system, i.e. a scaled-down version of the systems used by howitzers. The weapon's mass, while considerable, is simply not enough to make the recoil manageable without a fixed mount.

This all means that a fairly slight increase in cartridge size can translate to a lot more bulk and weight. The biggest anti-materiel rifles of this sort are mainly used against equipment and vehicles rather than personnel (hence the name) but can be used for long-range sniping. They cannot be fired on the move, however, which makes them no use in a situation where a sniper must fight his way out of a bad situation. Nor can such a heavy weapon be shifted rapidly from one position to another.

Thus the biggest rifles are deployed for carefully planned operations, such as supporting an operation against an insurgent stronghold. The sniper team will

Ammo Calibres

Most sniping weapons are cambered for calibres around 7.62mm (0.3in). Although there are exceptions, few use lighter rounds as they lack accurate range and stopping power. Heavier calibres are primarily used in long-range rifles or anti-materiel weapons.

| 5.56mm | 7.62mm | 8.58mm | 9.5mm |

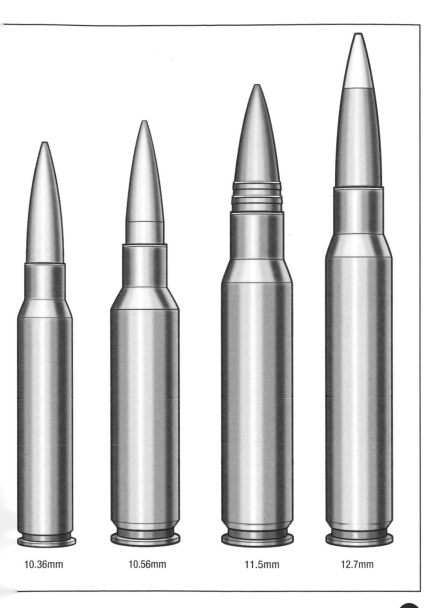

10.36mm   10.56mm   11.5mm   12.7mm

Barrett M82A1 Rifle

The first large calibre sniper rifle to achieve widespread popularity, the M82 has a large muzzle brake to reduce recoil. It fires an incredible heavy 12.7 x 99mm (.5 x 3.9in) cartridge, originally developed for the Browning M2HB heavy machine gun.

move from one firing position to another, as the operation is carried out, but they will require time to set up properly before shooting. A sniper with a 'lesser' weapon would be able to stop, take a suitable stance and shoot wherever he was, but a team lugging an anti-materiel rifle is likely

to be slower to respond to a new threat.

In some ways the very largest rifles can be considered to be in the same category as artillery. They take time to set up but can deliver effective supporting firepower once they are in position. There is most definitely a

place for such weapons, but for more normal sniping duties a lighter rifle is more generally useful.

Sights

It is possible to shoot accurately out to a very impressive distance using 'iron sights', i.e. the basic sights fitted to most rifles. However, this is not really desirable. Most snipers use a telescopic sight, which magnifies the target. Scopes are rated by their magnification. For example a 12x telescopic sight means 12-times-magnification; anything the scope is pointed at appears 12 times larger.

Types of Telescopic Sights

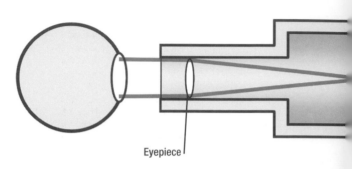

Eyepiece

Various types of telescopic sight exist. Almost all are adjustable for range and windage, and are extremely robust in order to retain their precise alignment despite the inevitable knocking and jarring they will encounter in the field.

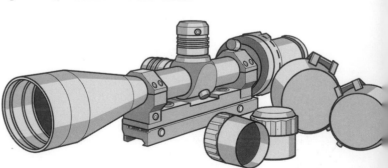

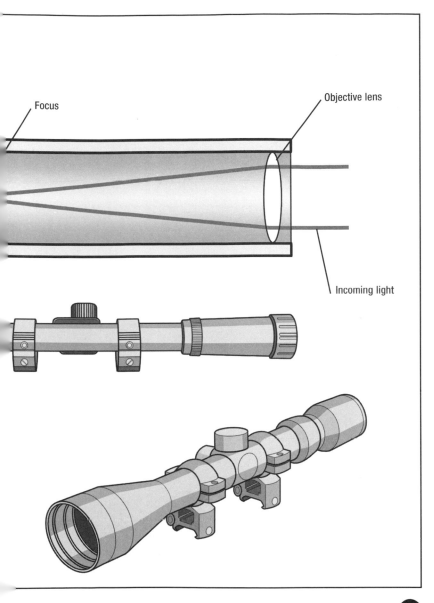

Focus

Objective lens

Incoming light

Aim Point

The correlation between sight alignment and bullet strike is indicated below. The upper line represents a telescopic sight with standard crosshairs, while the lower row represents a front blade sight.

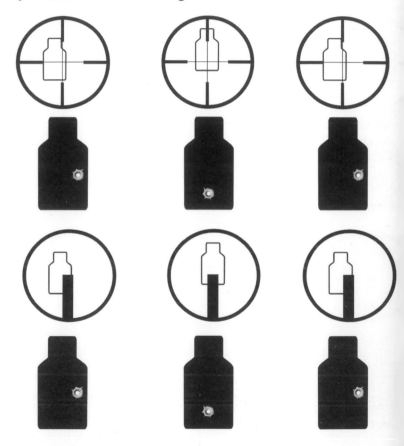

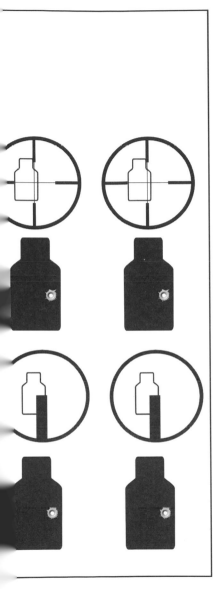

Depending on the light conditions, distance to the target and other factors, a sniper may use a telescopic sight with a magnification ranging from 4x to 12x or even more. A high-magnification scope is necessary when trying to shoot or spot a target at great ranges, but its field of view is very limited. Thus a 12x scope is excellent for examining a small distant area in great detail, while a 4x scope allows targets or threats in a wider area to be spotted.

When searching for concealed enemy personnel, a balance between field of vision and magnification is necessary. High magnification allows a close examination that may enable the sniper to spot well-concealed objects or people, but it may be that a movement that would have given away the enemy position may occur outside the sniper's field of vision. A lower-magnification scope may also be more useful when engaged at relatively short distances.

It takes considerable practice to develop an instinctive understanding of the relationship between scope magnification and movement of the aim point, i.e. how much the rifle must be moved to shift the aim point to a given target. It is possible to spot a suspect movement using the naked eye and then 'lose' the target once the scope is brought up to the eye. This is more common when using high-magnification sights, where a small movement of the rifle changes the aim point very little.

Being unable to quickly zoom in on a suspected enemy position is bad enough, but when trying to hit a moving target it is more serious. An unskilled shooter might become disorientated by the magnification of his sight and think he is moving his aim point much further than he actually is. Thus, although high-magnification scopes are extremely useful, they do have their limitations.

Some of these drawbacks can be overcome by practice and training, and a sniper would not be able to graduate from training if he could not effectively use a high-magnification scope, but others remain. A variable-magnification sight allows a suitable level of magnification to be selected, but this is not a perfect solution; the sniper must move to adjust his scope, which may give away his position or give the target time to move out of his field of view.

It is also often necessary to adjust a sight for range. The range post atop a telescopic sight allows the sniper to adjust his sights to compensate for bullet drop when firing at a distance. At short range, bullet drop is negligible; the round will hit more or less where the sights are pointed. However, the effect of gravity pulls the round down as it

Iron Sights

The term 'iron sights' refers to the simple sights traditionally fitted to rifles. These usually consist of a front post, sometimes with a loop, and a rear notch. Lining up the blade of the foresight over the target, when viewed through the rear notch, should ensure a hit. Iron sights are usually adjustable for distance, and for longer-range shooting some rifles have traditionally used a flip-up rear sight marked for distance in hundreds of metres. Using these raised sights effectively raises the muzzle and causes the bullet to arc further. Sniping is entirely possible with iron sights, but at long distances it can be hard to make out a target, and even if one can be seen, non-magnifying sights are insufficiently precise for long-range shooting.

travels, causing it to strike below the aim point or perhaps hit the ground before it reaches the target.

To compensate for this, it is necessary to shoot in a ballistic arc, with the muzzle of the rifle elevated. Greater range requires more elevation, which means that the sniper must either aim at a point above his target or adjust his sights so that putting the crosshairs on the target results in an arc that ends with the bullet striking the target. The range post allows the sniper to alter the alignment of his sights so that, with the scope pointed directly at the target, the rifle is actually angled upwards sufficiently to achieve the desired range.

Infrared and Thermal Systems

Advanced sighting systems using infrared (thermal) emissions or light-intensification technology tend to be more bulky than a standard optical sight, and require batteries, which have a finite life span. The decision to use these systems or not depends greatly on the mission at hand and the conditions likely to be encountered. They confer an enormous advantage when shooting at night, but are not helpful during the day unless the sniper is trying to see hostiles inside a darkened building.

A thermal imager detects the heat emanations from any object. Warm bodies such as people and animals,

the hot engine of a vehicle and so forth radiate more heat than colder ones. However, interpreting a thermal image requires some practice. Familiar shapes can be disguised by the patterns of warm and cold, and while a running man may be readily identifiable as such, picking an aim point on his body can be difficult.

Thermal imagers detect all heat sources, which can be confusing to the user. The temperature of the object is shown by its brightness, and two objects of the same temperature will have the same intensity. Heat sources other than people can confuse or even obscure the sniper's view of the target. A hostile who is on the far side of a small fire may be hard to distinguish, and even harder to target.

Thermal imagers are excellent reconnaissance tools. They can be used to spot otherwise invisible objects, such as hot pipes inside a wall, and will draw attention to objects and people that might otherwise be missed. They are not, however, infallible.

A cold vehicle will be the same colour and intensity as the cold wall behind it, and may actually be less obvious to the thermal imager than to an observer using just his eyes. Trying to navigate a dark area using a thermal imager can be disastrous, as many obstructions and trip hazards will be the same temperature as their surroundings

Night Vision Goggles

Night vision equipment is an important part of a sniper's arsenal. Goggles or binoculars allow a search to be undertaken without using the weapon's sight, and are highly useful when on the move in the dark. A weapon-mounted sight is necessary for shooting in the dark, however.

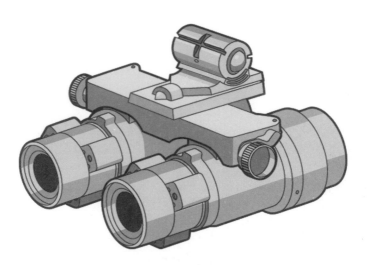

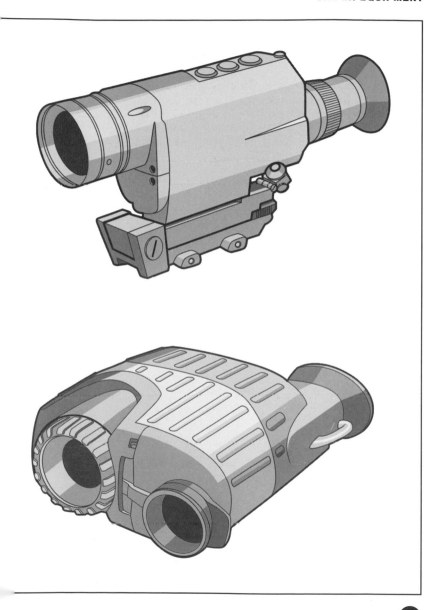

Types of Supports

Sandbags and sand socks both offer the sniper infinite variability. The aim point can be microscopically altered by pressure on the sand container, which cradles the weapon rather than supporting it at a single point.

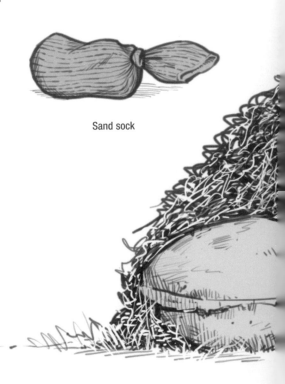

Sand sock

and therefore all but invisible. Nevertheless, a thermal imager may give the sniper additional capabilities necessary for a given mission.

Starlight Scope

A low-light or light-intensification sight does not rely on thermal emissions but electronically amplifies whatever light is available. A device

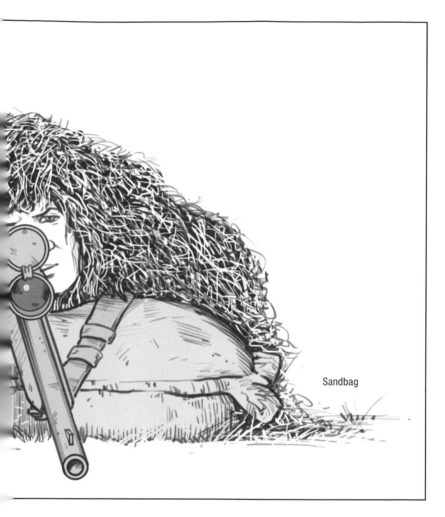

Sandbag

of this sort is sometimes called a 'starlight scope' as it can provide a useable image using minimal light such as a moonless sky. A starlight scope is no use in complete darkness; it cannot see into a cave or blacked-out building. However, it does allow shooting in conditions where it might otherwise be difficult even to find a target.

The Sand Sock – A Sniper's Friend

Instead of supporting the weapon directly with the lead hand, a sniper can instead use a sock that is a little more than half full of sand. The weapon rests on the sand, and height can be adjusted with infinite precision by squeezing the sock. Alternatively, the weapon can be rested on the sniper's fist, which is in turn atop a supporting object. Tightening or loosening the fist will raise or lower the aim point of the weapon.

Field Expedient Tripods

A weapon support can be quickly improvised by fastening together two or three sticks, or using a forked stake. The latter requires that a suitable stick be located and trimmed, whereas a bipod or tripod can be created from almost any length of wood chopped into convenient lengths.

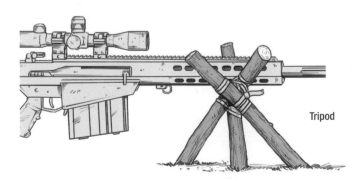

Tripod

Laser Technology

Snipers are trained to judge distance by eye, but a laser rangefinder is an invaluable aid to finding a precise distance. Even a slight error in estimating distance can result in the round falling short or flying over the target, or a moving target not arriving at the aim point at the same time as the bullet.

Some snipers also carry laser designators, allowing them to 'paint' or 'lase' targets for aircraft or artillery fire.

Bipods and Monopods

Some sniper rifles have a built-in bipod or monopod. The latter is a single post that can be used to support the weapon. While less stable than a bipod, a monopod is also lighter and less likely to snag on obstructions while the sniper is moving.

Pushing a rifle with a bipod under the barrel through a small opening in hard cover or heavy undergrowth can be impossible; a monopod does not add to the width of the weapon and

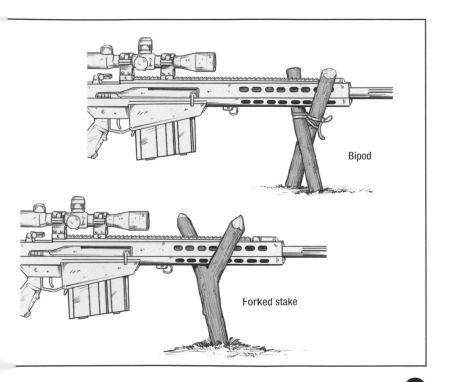

Bipod

Forked stake

is less likely to drag undergrowth about as the sniper tries to move or extricate his weapon. The sudden movement of a bush just after a sniper shot may be all it takes to attract an immense volume of return fire.

Suppressors

Many snipers make use of a suppressor to reduce the noise of their weapon by trapping some of the propellant gases and dissipating their energy through a series of internal baffles. There is in truth no such thing as a 'silencer', as all firearms make at least some sound, but a suppressor greatly reduces this. Suppressed weapons do not guarantee that a shot will not be heard, but they do make it much harder to pinpoint the shooter's position. The suppressor also eliminates muzzle flash and any tendency to kick up dust around the weapon's muzzle, both of which can give away a sniper's position.

A suppressor is used with subsonic ammunition, which travels more slowly than conventional rounds. This reduces accuracy and alters the ballistic performance of the weapon while the suppressor alters the balance of the weapon. The effects are more noticeable at longer ranges, but as a rule a sniper shooting at long range has less need of a suppressor. Suppressed weapons are mainly used in covert operations or when a sniper is operating independently, especially if enemy forces are patrolling or are otherwise active in the area. When

Overpenetration

A hard, pointed bullet may penetrate body armour and cover effectively but may also pass right through flesh without slowing down, taking much of its energy with it. This does not matter if the round hits a vital organ, but in the case of a marginal hit, overpenetration may result in a non-disabling wound. The bullet may continue to be potentially lethal on the far side of the target, which can be a problem in some situations. Achieving a balance between penetration to defeat cover and armour, and lethality upon striking flesh, is a tricky business.

shooting in support of an infantry force directly engaged with the enemy, stealth is less important – the hostiles already know that there is a firefight going on, and the noise of the combat will probably mask that of the sniper's rifle.

Ammunition

Ammunition is extremely important to a sniper. Modern rifle rounds, unless very cheaply made, tend to possess fairly uniform ballistic characteristics, which means that any given shot will probably perform very much like any other from that batch of ammunition. However, 'fairly uniform' may not be enough. Ideally, a sniper wants ammunition that will perform in exactly the same way with each shot. He has learned to shoot consistently and has a rifle that will do so too; even a small variation in ammunition performance can negate all that.

Snipers use 'match grade' ammunition, which is produced to extremely high standards. The very best is known as 'green spot' ammunition, because it is marked with a green spot. This shows that the rounds are among the first 5000 out of a new mould, which will therefore not have picked up any small dents or irregularities that could alter the shape of a bullet. Even a small irregularity alters airflow and weight distribution as the bullet spins on its way to the target.

Most sniper shots are made with standard anti-personnel ammunition, which is constructed for a good balance between stability in flight, range, penetration and wounding properties. The standard round is known as military ball ammunition, but other types may be used instead. Tracer rounds have a small amount of incendiary material in the base, which provides a bright light for tracking a shot or marking a target for other troops. Of course, tracers will give away a sniper's position, so they are used only when the circumstances merit the risk.

Armour-piercing Ammunition

Military snipers also use various types of armour-piecing ammunition. As a rule, armour-piercing rounds penetrate cover and body armour better than standard rounds, but have inferior wounding properties. They are used against protected personnel but can also damage equipment.

Large anti-materiel rifles can make use of high-explosive armour-piercing rounds, which cause severe damage to whatever they hit. Although there has been some controversy about using explosive rounds of this type against personnel targets, they have been ruled to be legal for all uses by the US military. However, 'ball' rounds are generally preferred for personnel targets, not least because anything more potent represents expensive overkill.

Putting a bullet where you intend it to go is a complex business. There are two key components: the ability to decide where to shoot, and the ability to shoot consistently. If either is lacking, then the sniper will be ineffective.

Sights are of enormous assistance, especially at long range, but they are useless without the skill to use them. To hit a given target, the sniper might have to aim at a point that is a significant distance from the target's current position. Putting the crosshairs directly on the target is only viable when it is not moving, at fairly short range with little wind. The shooter must be able to pick a suitable aim point, taking into account the movement of the target, time of flight of the bullet, wind, bullet drop and all manner of other factors, and it is here that he aims.

If the sniper can always create exactly the same relationship between his eye, the sight and the muzzle of the weapon, then he will shoot consistently, providing the weapon is kept perfectly steady. This gives the sniper the ability to hit his

. . . .

Snipers have to ensure that the relationship between the sniper's eye and his sight is always exactly the same, and that his weapon is correctly lined up with his body. If both are achieved, the round will go where the sniper is aiming it.

6

A sniper must be able to shoot accurately from a variety of positions.

Marksmanship

Hawkins Position

In the Hawkins firing position, the weapon's stock is under the sniper's shoulder rather than against it. The weapon is

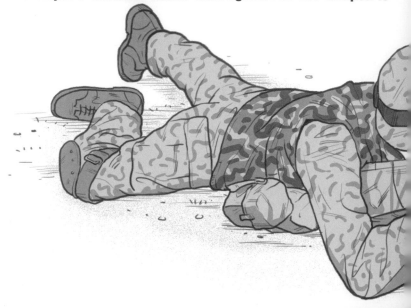

Stock Weld and Eye Relief

A good firing position requires a close 'weld' between cheek and stock, and it is essential that the cheek is in exactly the same place each time a shot is made to ensure the correct eye relief is maintained. This is the distance between the eye and the rear of the sights and is essential to avoiding eye strain when using sights for an extended period.

supported by the sniper's fist, gripping the front sling swivel, with his lead arm straight to help absorb recoil.

chosen aim point; and if it is chosen correctly, he will hit the target. So, the skilled sniper must be able to predict where his target will be when the bullet arrives, select an aim point that will cause the bullet to pass through his chosen aim point, and then consistently hit that aim point.

Some elements of a shot can be set up in advance. A sniper rarely takes position and just starts blasting away at the enemy. He may choose

to pass up a target that presents itself if he is not ready to shoot. Making ready requires taking the time to estimate range and wind speed, atmospheric conditions and the like, as well as factors unrelated to the shot itself but vital to the sniper's survival, such as the presence of enemy patrols, guards and observers.

Armed with a solid feeling for local conditions and settled into a steady firing position, the sniper can then

Using Loopholes

A loophole broken in a wall allows the sniper to see and shoot out of the building, but may cause suspicion or even attract fire. Covering the loophole with a semi-opaque material can make it less obvious to enemy observers, at least until the shooting starts.

begin to set up his shot. The final phase is short; the sniper takes a couple of breaths to oxygenate his blood, lets half a breath out and holds the rest, and takes up the slack on the trigger. His rifle comes to aim

at the chosen point as he does this, settling into its final precise alignment as the sniper holds his breath. He now has a few seconds, at most, to decide whether to shoot or wait for a better opportunity.

Improved Infantry Sights

Increasingly, the rifles of ordinary infantrymen are equipped with improved sights, with the traditional iron sights serving as a backup. Improved sights include reflex or red-dot sights, which project an aiming point on to a clear lens. The dot indicates where the weapon is aimed, even if the user moves his head. This allows more rapid target acquisition during a close-range firefight, offering some of the speed advantages of hip-shooting along with improved accuracy.

ACOG and SUSAT

Many infantry rifles are now fitted with low-power magnifying sights for longer-range combat. The British Sight Unit Small Arms Trilux (SUSAT) is a fixed-power 4x sight, while the Advanced Combat Optical Gunsight (ACOG) is available in a range of models from 1.5x up to 6x. These sights are fixed-power, i.e. they cannot be adjusted. Nor are they intended for use by snipers, although one could be pressed into service at need. These magnifying sights allow infantry soldiers to shoot accurately at longer distances, improving the marksmanship of the force as a whole.

Adjustable Triggers

Every sniper has his own preference as to how much trigger pressure should be required. Adjustable triggers allow the sniper to set the required pressure exactly as he wants it. If the trigger pull is too heavy this is likely to reduce accuracy; too light and it may be difficult to judge when the weapon will fire. A good sniping weapon has a light but crisp 'break'. Two-stage triggers are common, with the first stage 'setting' the trigger ready to fire with just a tiny increase in pressure.

L85 (SA80)

If the sniper does decide to shoot, he requires only the tiniest increase in pressure on the trigger. As a result of intimate familiarity with his rifle, he can bring it to just short of the point of firing and hold there until the moment is right. Then the trigger is pulled straight back with the tip of the finger, triggering the shot with an absolute minimum of movement.

This act is the culmination of possibly hours or even days of stealthy movement, observation and

Firing Position

A good shooting position is all about making a perfect 'weld' between cheek and stock. In this case, the left hand supports the stock and pulls it into the sniper's shoulder for a tight, snug fit.

waiting. It is all-or-nothing; if any aspect of the shot is wrong, then the bullet will miss its target. All of the sniper's skills are used to get him to the point where he can make ready the shot, set up the shot and, finally, take it. Of course, not every shot hits its target, but the sniper aspires to perfection – one shot, one kill.

Shooting Positions

Snipers shoot from a range of positions, many of which look to be

The weapon lies along the natural axis of the sniper's body; the aim point is adjusted by moving the sniper, not just the weapon.

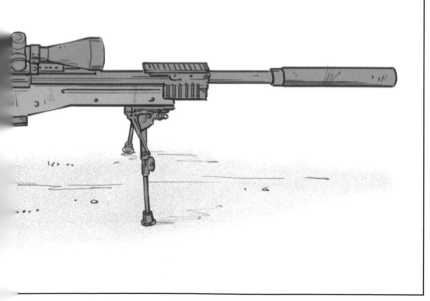

quite uncomfortable. All are designed to perform the same critical task – to keep the rifle steady and the relationship between the shooter's eye and his sights the same.

The rifle must be well supported and pulled in firmly but not too hard to the shoulder, with a good solid 'weld' between the shooter's cheek and the stock. Pulling the weapon into the shoulder too hard or gripping it too firmly will reduce stability rather than increase it.

The rifle is supported and held in place by the sniper's trigger hand, his cheek and his shoulder, all of which will be in the same place each time. His other hand can take many positions, either gripping the weapon or acting as a passive rest. Often, both hands are close together, with

the lead hand supporting the rifle close to the magazine or just in front of the trigger assembly. This ensures that there are no lateral forces on the barrel; the sniper supports the weapon as a whole by the receiver, and the barrel effectively 'floats'.

Any firing position the sniper may adopt must allow the aim point to be adjusted without excessive movement, and must avoid tension in his body. A shooter who is fighting his own body will not hit anything. Movement may be fairly limited, but this is not a major issue. Any target that would require a large, rapid movement to acquire would normally be passed up because, first, a hit would be unlikely and, second, snipers do not like to make rapid movements in case they give away their position. Judging which shots

Prone Shooting Position 1

A classic prone shooting position minimises the sniper's silhouette and provides maximum contact with the ground for stability. Recoil forces are aligned straight back along the sniper's body axis, minimising the effects of firing even a large, powerful anti-materiel rifle.

Bone Support

A sniper's weapon should be supported by his bones rather than his muscles. The structure of his body should be aligned so that he can relax his muscles without altering the position of the weapon. Relaxing reduces tension in the body and makes any shot more precise.

Prone Shooting Position 2

An unsupported prone shooting position requires that the sniper's left arm support the front of the weapon, rather than using a rest or a bipod. The rifle is cradled between fingers and thumb of the forward hand, but is supported rather than gripped.

not to try for is an important ability, and a skilled sniper will prefer to wait for a good clean shot rather than snatch a low-percentage one and perhaps waste his opportunity.

Prone Position

A sniper's shooting position may have to be maintained for some time. A certain amount of discomfort can be ignored, but unnatural positions cannot be held for long. Certain postures lend themselves ideally to

sniping, and of all of them a prone firing position is ideal for many purposes. It allows the rifle to be easily supported and requires minimal effort to maintain. A prone sniper is also unlikely to be spotted, because he will not be outlined against whatever is behind him. A prone position behind an obstruction such as an earth bank is excellent, as it offers support, concealment and cover from return fire. A prone position might not be possible in the

case where a sniper is concealed inside a building, but it may be possible to rearrange the furniture to create a suitable firing position.

A large table dragged into the middle of a room allows the sniper to lie prone and still see out of the window. It may also be possible to create 'loopholes' in a building by removing parts of a wall or roof tiles, or to find an elevated position that allows the sniper to shoot over the edge of a roof.

On One Knee
The other 'classic' firing position is kneeling on one knee, with the elbow of the front hand supported on the front knee. This position is extremely stable and offers improved all-round vision compared to lying prone. Kneeling allows the sniper to stand up quickly and move off if necessary, or turn around to look for targets in another sector. This may be important when on the move; the sniper can keel to shoot, and move on. Of course, a kneeling sniper is easier to spot and hit than one who is lying down.

Kneeling allows the sniper to see over obstructions and rest his weapon on a solid object. Height is an important consideration here; if the object is of an awkward height, then it will not make a good weapon rest. The sniper cannot afford to have to perform contortions in order to see through his sights, so he may have to

Kneeling Shooting Position

The key to a solid kneeling position is to have the front leg as vertical as possible, while the rear heel is as close to the spine as possible. The sniper essentially sits on his rear heel with his body aligned at 45 degrees to the target.

Natural Point of Aim

The natural point of aim is the position where the alignment of the sniper's body naturally targets his weapon on the desired aim point. An unnatural position requires effort to maintain, which can result in a miss. For precise shooting, the sniper aims with his entire body. His weapon becomes an extension of his body and lines up naturally.

pass up an otherwise good position. However, it is possible for a sniper to kneel beside an object and rest either the weapon or his shoulder against it. A kneeling position is useful for shooting around, as well as shooting over, large objects.

Standing

Most of the comments applying to kneeling are also applicable to a standing position. Snipers will normally shoot from a standing position only when behind cover, or when firing out of a building. The sniper may adopt the classic rifle marksman's position, body sideways on to the target and lead hand well forward on the weapon's forearm. His hand may be close to the trigger assembly, as with many other positions, but a standing posture is inherently less stable than many others. As a rule, the closer the rifle is to its anchor points on

the ground or any object used as a rest, the more stable it will be. A standing position places the rifle near the top of a fairly large structure (the sniper's body and weapon), and holding it with both hands close together may amplify any sway that results. A traditional long grip is usually more appropriate than keeping both hands close together.

Sitting

Shooting from a sitting position may seem counter-intuitive, but it does provide a stable base to support the rifle. The sniper's buttocks and feet form a tripod, and he may be able to brace himself against a solid object for an additional anchor point. A seated position is low but affords better all-round observation than lying prone. It is also relatively easy to maintain this position for an extended time.

Standing Shooting Position

Wherever possible, the weapon is rested on a wall or other support. The sniper can adjust his height to that of the wall by standing on something or by crouching down, but there is a limit to how far he can go before his position becomes difficult to maintain. Tension in the body from an awkward position will offset any advantage gained from resting the weapon.

A variant of the seated position is to lie back against a supporting object with the rifle supported on a raised knee. While apparently better suited to a siesta than sniping, this position is low and stable. It can be used to observe or shoot over low cover, or to remain unobtrusively on

Seated Shooting Position 1

A seated shooting position is in some ways a variant of the kneeling position. The front knee supports the front arm and thus the weapon, while the sniper's weight rests on his rear leg. This is a very compact position that minimises the sniper's profile.

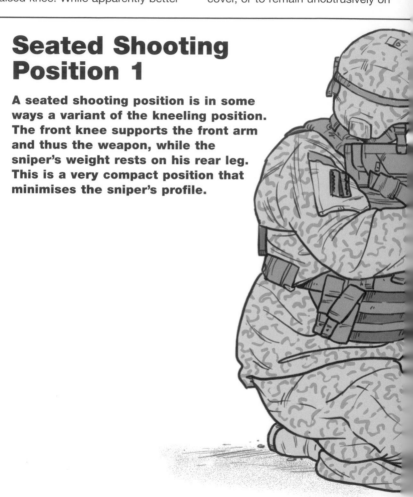

the alert during a long waiting period. A similar position is used to good effect by some long-range pistol-shooting competitors, who use maximum contact with the ground to minimize movement in their bodies while aiming a handgun at an extremely distant target.

Basic Principles

There is almost infinite variation in exactly how a sniper positions himself to shoot, but the basic principles are the same. The weapon must be steady and this requires good contact with the sniper's body as well as minimal movement within his body. This, in turn, necessitates a good solid contact with the ground and a stance that minimizes possible movement.

Shooting positions are adapted to

Seated Shooting Position 2

A supported seated position uses any available support. Sitting down lowers the sniper behind his cover or rest and allows the sniper to remain in position, yet ready to move on at an instant's notice, for an extended period.

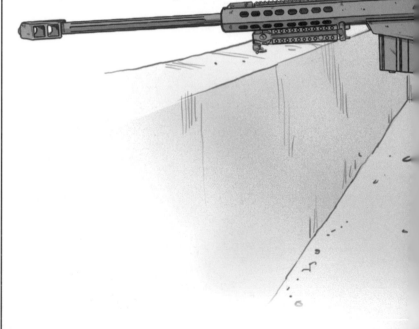

the local environment. For example, it is possible to adapt a standing position by putting one foot on a fairly high object and resting the front elbow on the raised knee for added stability. Such a position might be used in a cluttered environment or when firing over a wall.

The weapon can be rested on any suitable solid object, such as a wall, vehicle or large rock. However, this often requires that the sniper adjust his

position to suit. If he must bend his back or legs too much to accommodate the resting position, his posture will become awkward and difficult to hold. He might be better in such a case to either move to the end of the object and brace against its side, or to move back from it and fire over it from a standing or kneeling position. The object can still serve as cover and concealment, even if it proves unsuitable for use as a rest.

Environmental Conditions

A sniper must be able to account for a wide variety of conditions, some of which would not occur to the average person or even most experienced shooters. The effects of

Range Card

Preparing a range card is one of the sniper team's first tasks after moving into position. Sketching the terrain on to the card ahead of time allows the sniper to look up the distance to a given target or reference point, rather than having to estimate it on the fly.

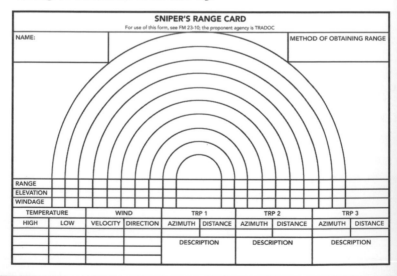

gravity on a bullet are one such factor. The term 'point blank' has come to be associated with an extremely close range, but in fact point-blank range can be quite a long way. The term originated with artillery gunners, signifying the outer range at which the gun would hit a target that it was aimed at directly. Beyond point blank range, the gun had to be elevated.

Obviously, as soon as a bullet leaves the muzzle of a rifle it begins to be pulled down by gravity, but for a high-velocity firearm, point-blank range is a lot further than the couple

Using a Sniper Scope

Sniper sights are either of a single post (top) or crosshairs (lower) type. The tiniest variation in the angle that the sniper looks through the sight affects the fall of the shot. This diagram shows the relationship between changes in the angle of viewing and the landing of the round.

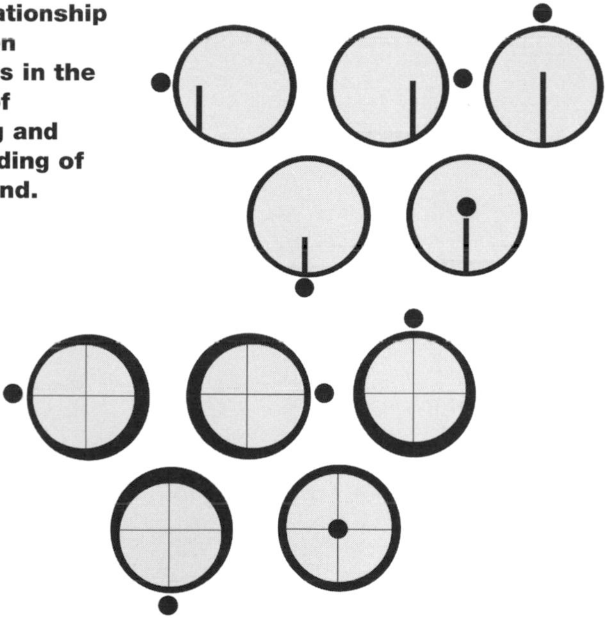

Ballistic Arc

Long-distance shots must travel in a high arc in order to hit the target, which can be a problem if there are intervening overhead obstacles. A higher muzzle velocity equates to less bullet drop at any given range, allowing a

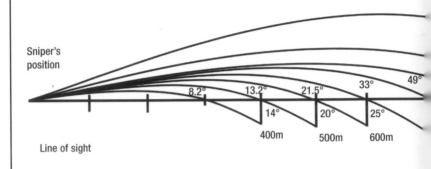

- **Illustrates highest point of flight from 300 metres to 1000 metres in 100 metre increments.**

- **Shows bullet drop for each succeding 100 metres.**

of metres implied by common usage of the term. Indeed, much infantry combat is conducted at 'point-and-shoot' ranges, when firing from the hip or on the move has a reasonable chance of success. At such relatively close ranges, being a little high or a little low may not have

flatter and more direct trajectory at short ranges and a greater maximum effective range.

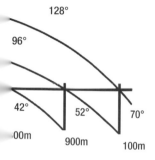

128°

96°

42°   52°   70°

00m   900m   100m

much effect, but when firing at several hundred metres there is a strong possibility that the bullet will ly over the target or fall short.

Longer Range

For shots at longer ranges, bullet drop must be compensated for. This normally means adjusting the weapon's sights so that when they are positioned on the target, the barrel is actually elevated. A bullet that leaves the muzzle travelling slightly upwards will be pulled down by gravity and should be at the necessary height when it reaches the target. This, of course, assumes that the weapon's sights are correctly adjusted for the range, which in turn requires the sniper to have estimated the distance to the target correctly.

Even if the weapon's sights are perfectly aligned on the target, if the range estimate is off, then a long-range shot will miss. Thus snipers must learn to estimate distances by eye. Tools such as laser rangefinders are more accurate than any human can be, but they are not always available. Snipers must learn to use their own abilities first and foremost, and not to rely on artificial aids.

Wind also has a profound effect on a long-range shot. A small bullet, travelling fast, will not be deflected very much during a short flight to a close target. However, a longer shot can be sufficiently affected by even a light wind that the target may not even realize he is being fired upon. A sniper must be able to judge the effects of wind on a shot, which in turn means that he must be able to estimate the strength of the wind and

Ballistic Arc and Retained Velocity

The ballistic path of a bullet is not perfectly symmetrical; drag from the air slows the round as it flies, causing its descent to be steeper than its rise. The retained velocity of the round when it reaches the target will thus be less than its muzzle velocity. Wind effects can cause side-to-side departure from the line of sight, and can increase or decrease the steepness of the bullet's arc if blowing towards or away from the target.

its direction. Developing the skill to do this, and then to figure the data into his preparations for a shot, is an essential part of the sniper's training.

On a very long shot, wind calculations become highly complex. The strength and even the direction of wind can vary over the distance covered by a long shot, and there may also be eddies around the corners of buildings or other objects. Air temperature and humidity also affect a shot by changing the amount of drag the air exerts on the bullet as it passes through. More drag means that the round will lose energy sooner, falling a little shorter. Air temperature can also create optical effects that make a target seem closer or further away.

A sniper needs to be able to estimate the aggregate effect of these many factors on his shot, at the precise moment he squeezes the trigger. The only way to learn to do

this is from experience. The sniper must make many, many long shots under varying conditions, with a spotter to tell him where the round actually went. This data is recorded on a card, allowing the sniper to gradually build up a feeling for the many factors that might affect his bullet in flight.

All snipers are taught to estimate wind speed from the movement of vegetation or litter, and how this will affect a bullet in flight. Yet, even with all this knowledge, picking an aim point is something of a 'black art', with some snipers being instinctively better at it than others, sometimes without really understanding why they can fit all the pieces together to create a perfect shot.

Rifle Characteristics
Most civilian target shooters and combat troops will fire repeated shots from their weapon, which

Leading, Tracking and Trapping

Leading a target simply means placing the aim point just ahead of the target's current position to account for its movement. Tracking means following the movement of a target with the sights, which can also require leading somewhat. Trapping (or ambushing) is the act of placing an aim point somewhere that the target is likely to move through, and shooting when he reaches it. An erratically moving target can be tracked in the hope that it either stops or becomes predictable. If the target stops moving, the sniper stops tracking and holds his aim point, then shoots. If the movement becomes steady, the sniper can lead the target a little then shoot.

Sighting a Moving Target

Shooting directly at a moving target is an exercise in futility at any but the closest of ranges. The target must be 'led' a little to ensure that bullet and target arrive at the same spot simultaneously.

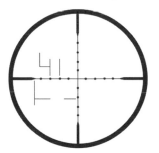 

Urban Cover

Urban areas offer a wide range of possible cover and concealment. Most building materials, e.g. bricks, are reasonably resistant to enemy fire but can be penetrated by a powerful round. Secondary fragmentation from a bullet striking a wall is also a hazard.

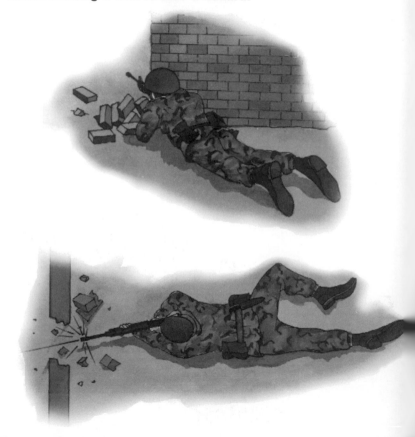

causes a small amount of fouling due to unburned propellant. They will also be firing from a hot barrel most of the time. The effects of these factors on accuracy are small, but significant over a long distance. A sniper will often fire only one shot, and will thus shoot from a clean, cold barrel. He needs to be aware of the characteristics his rifle possesses with a cold clean barrel and how they will change if he has to take several shots. Like other skills, this is a matter of experience gained in many, many shots.

Moving and Protected Targets

Shooting at a moving target is never easy, even when its movement is predictable. A target that changes speed and direction, which disappears behind intervening objects or ducks into cover and comes out again unpredictably, may be impossible to hit. This is one method of defeating enemy snipers: to break cover suddenly, rush or roll to the next hiding place and vanish again. This sort of movement is exhausting, however, and is normally used only by personnel who expect to come under fire.

Urban Terrain

Well-trained troops will minimize their vulnerability to snipers on a constant basis. British forces deployed to Northern Ireland learned how to use the urban terrain to make themselves a hard target while on patrol on the streets. Walls, postboxes and doorways all provided cover, and troops habitually positioned themselves so that such cover gave protection from at least some directions. There was often no way of knowing whether or not a sniper was active in the area at the time, but habitual anti-sniper techniques undoubtedly saved a lot of soldiers' lives.

The troops of many nations use the same concepts. Inexperienced or complacent soldiers might stand around in the middle of the street when a patrol stops moving, and of course it might be necessary to be in the open, but, as a general rule, well-trained troops will remain close to hard cover and make use of it whenever they stop. Something as simple as dropping to one knee will make a soldier harder to hit and may cause a sniper to pass up a shot.

In the Bunker

Yet snipers must be able to hit targets that move and make good use of cover. In the case where a hostile has taken cover or lain down to minimize his target size, the ability to shoot precisely and accurately can enable the sniper to hit him anyway. The same applies to hostiles firing through a very small gap in hard cover, such as out of a bunker's firing slit. Protected by solid concrete, an

Shooting Through Cover

It is sometimes possible to estimate a target's position from as little as a boot or a rifle barrel sticking out of a bush. This may allow a sniper to take a best guess and shoot, but it might be a decoy intended to draw him out. Experienced snipers are unlikely to fall for any but the most convincing ruse, but a relative beginner might.

Practice Target Boards

Practice targets resemble enemy personnel for two reasons. One is to train the sniper to place his shots on a human target, with realistic reference points. The other is to ensure that he is used to shooting at human targets. By making training as realistic as possible, the chances that he will hesitate are diminished.

Lock Time

The firing mechanism of early gunpowder weapons was known as a lock. There was a significant delay between pulling the trigger and the weapon discharging, especially with flintlock and matchlock weapons. This delay is still known as lock time, though today it is usually very short. Lock time cannot be reduced to zero; the firing pin must be released and travel a short distance to strike the primer, which then ignites the cartridge. If the rifle moves during this delay period, the shot will miss.

enemy machine-gunner might be all but invulnerable to return fire from most infantrymen, but a sniper should be able to put a round through the firing slit. Even if he cannot see the gunner inside a dark bunker, he can determine his position from the orientation of the weapon and shoot at where he must be. During World War II, snipers were often called upon to deal with bunkers in this manner.

Clues and Guesswork

It is also possible to shoot through some of the things that an enemy might hide behind. Hard cover will stop a bullet, but most forms of cover offer concealment more than protection. It may be necessary to estimate the hostile's position from clues such as a rifle barrel poking out of a bush or from behind a door.

Snipers are rarely keen to rely on guesswork, but sometimes an educated guess is all that is needed.

If an enemy ducks out of sight behind a clump of vegetation, there are few places he could be. A combination of psychology and common sense will tell the sniper where the hostile is likely to position himself, and if it is important that a shot is taken, he can take it with some confidence, even if he cannot see the target. This is by no means an ideal situation, but there is still a reasonable chance of success.

Snipers have to be skilled at 'filling in' a partial picture. A car door, for example, will not stop a bullet, but it will conceal whatever parts of the hostile are behind it. This can affect the accuracy of many shooters – even though they can see parts of the hostile above and below the door, it becomes difficult for most people to aim accurately at him. A sniper must be able to overcome this tendency, in case he has to shoot at a partially concealed target.

Ways of Shooting

There are essentially two ways to hit a moving target, other than by spraying the general area with automatic fire and hoping for the best. The first is to pick an aim point ahead of the moving target and shoot just before the target reaches it. The sniper must time his shot perfectly, taking into account the tiny delay between the decision to shoot and squeezing the trigger, the 'lock time' of the weapon (i.e. the small delay between trigger pull and the round being fired) and flight time of the bullet. If he is right, bullet and target will reach the aim point at the same instant.

The second method is commonly used by clay-pigeon shooters. The sniper tracks the target's movement, starting behind him, and moves his aim point just a little faster than the target. As the crosshairs move ahead of the target, the sniper shoots. The series of small delays mentioned above give the target time to move into the aim point.

Both of these methods are valid, and may be used under different circumstances or according to the sniper's preference. The former is more likely to work with an erratically moving target than the latter; it is hard to track someone who changes speed and direction.

Trajectory and Line of Sight

A line of sight is, by definition, a straight line from the sniper's eye to the target, through his sights or scope. His bullet, however, will travel in a ballistic trajectory, flying in an arc as it departs the weapon slightly upwards, then falling back to earth due to gravity. Higher velocity results in a flatter trajectory, and at short range the trajectory may be more or less flat. Some targets cannot be hit due to an overhead obstruction somewhere on the bullet's trajectory, even though the sniper has a clear line of sight.

Avoiding a Clearing

While it may be quicker to move across a clearing, a sniper team will go around it. Not only are they more likely to be seen crossing open ground, but hostiles in the clearing may spot them if they approach too closely. The team will instead remain in the denser woods where they are better concealed, even if this means moving quite slowly.

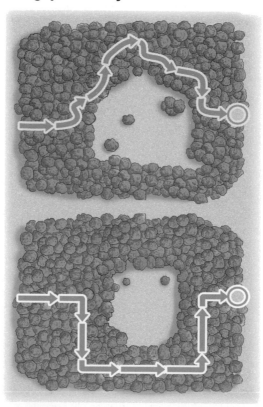

Moving Target

One way to overcome, to an extent, the difficulty of hitting a moving target is to predict when and where it will stop. This requires a quick appraisal of the situation and the application of logic. A good sniper can identify positions of cover that a moving hostile will duck into; or a predictable pattern of movement such as an enemy gunman who fires from one window of a building and then moves to another. This may cause some of his enemies to lose track of him or be unable to predict where he will reappear, but the concealed sniper, who is not under fire himself, has time to watch and analyze the pattern.

Sometimes it is obvious where an enemy is going to stop. A senior officer walking towards a helicopter or vehicle is likely to be planning to enter it. He may have to halt while a door is opened for him, or will at least move more slowly as he clambers aboard. If he is walking towards a communications set, then he is probably going to use it, which will mean he will become stationary.

Other personnel will also move predictably at times. A hostile who is wildly dodging across an open space will probably be heading for a position of cover and will stop when he gets there. Depending on where he thinks the enemy is located, he may be entirely exposed to the sniper's fire while he takes shelter from the perceived threat. The sniper

need only wait until he gets there.

The same applies for troops headed for sandbagged positions or emplaced machine guns. Chances are that they are going to use the weapon or at least avail themselves of the hard cover on offer. The sniper does not need to make a hasty shot at a moving target in this case. Instead, he waits for the hostile to stop, or to position himself, at the weapon. Once he is stationary, he will be a much easier target, even if he has partial cover. If the hostile stays hunkered down under cover, then he is effectively out of the fight and the sniper can seek another target. If he exposes part of himself to shoot, the sniper can eliminate him.

Weighing it Up

Thus sniping is not just about being able to shoot straight. The sniper must be able to think through his shot, weigh up a range of factors and, in some cases, predict what the target is about to do. A soldier who can do all of these things, and has the necessary observation and stealth skills to get himself into position to make the shot, will be able to complete sniper training and graduate. Those who have some but not all of these skills will return to their units with new knowledge and will become more effective infantrymen. They may even find themselves showing their comrades how not to become a sniper's victim.

Although today's snipers are aided by a range of highly advanced electronic equipment, many still rely on the decades-old technology of the bolt-action rifle. Technology can assist, but in the final analysis it is the skill of the sniper with his rifle that makes all the difference. Here, a British Army sniper team provide overwatch support for comrades in Afghanistan with an L96 sniper rifle.

PART THREE:
Snipers In Action

As already noted, snipers rarely operate alone. They are usually deployed in loose proximity to friendly forces, and function best with the support of an observer. The sniper team may also be assigned a security element, although this can be counterproductive if it makes the team too large. The sniper team and security element will separate before the team approaches its firing position. Even when several sniper teams are deployed to an area, they will disperse and avoid moving together.

If the sniper is deployed in a static location or as protection for a vehicle convoy, he may well be integrated with the rest of the force. In this situation the sniper's exact position may be unclear to hostiles, but the presence of a fairly large group of combatants is unlikely to be overlooked. On the other hand, if the sniper team is deployed in the field or must move into hostile territory, a team of two or perhaps three men has a reasonable chance of avoiding detection, whereas a large force would be all but impossible to conceal.

. . . .

Modern snipers tend to function in two- or three-man teams that include a spotter/observer, and often a security man to watch the team's back and look out for threats.

7

Snipers are part of a larger operation or organisation, applying their specialist skills towards a common goal.

Sniper Operations

Primary and Other Weapons

The sniper's primary weapon is his precision rifle, but he will often carry a handgun or a personal defence weapon, such as a small sub-machine gun, for emergency use. The observer may have a sniping weapon of his own, or a semi-automatic marksman rifle, which can lay down fairly rapid fire. He may instead carry an assault rifle. There are arguments both ways. On the one hand a second precision weapon effectively adds another sniper to the force. However, the observer's function is different from that of the sniper, and if he is shooting, he is not carrying out his primary function.

For those sniper teams that do not include a third man as a security element, the observer's weapon may be the only way out of a situation that has gone badly wrong. The logic here is simple: sniping is efficiently accomplished by the specialist sniper

L115A3

A bolt-action weapon, the British Army L115A3 uses heavy 8.59mm (0.338in) Lapua Magnum ammunition which has a greater effective range than a 7.62mm (0.3in) round. The rifle has an accurate range of 1500m (1640 yards), but has been used to hit targets further away in some instances.

and his supporting observer, and if a close-quarters emergency transpires, then the observer provides heavy firepower to deal with the situation or facilitate an escape.

Policies regarding the armament and composition of a sniper team vary between forces. It is, however, common to designate one team member as the sniper, who is the leader of the team. The observer and any security element that may be included in the team are subordinate, even if all team members are extremely experienced and qualified snipers. Clear leadership is necessary for the smooth functioning of the team and to avoid potentially fatal mistakes.

The sniper is in charge and makes critical decisions, such as whether or not to shoot and which targets to select. He also normally makes the shot. However, it is common for sniper and observer to swap roles as shooter and observer during a long

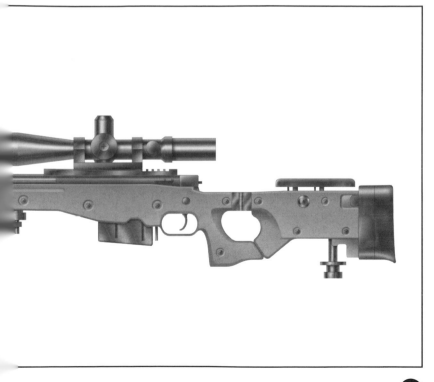

stalk or static deployment in order to avoid fatigue. The designated sniper is still in charge, even if his observer is currently acting as 'shooter'.

Team Effort

Close co-operation is vital; sniping is a team effort rather than a matter of orders given and obeyed. The sniper and observer may have to communicate by way of hand signals or tapping one another, and an experienced team who thoroughly know their business develop an instinctive understanding of what each other will be thinking and doing. If both sniper and observer properly understand sniping, then something as simple as looking at one of a group of potential targets will indicate intent. Indeed, a good observer will often know which target the sniper will select next, because he would do the same.

Role of the Observer

While the designated sniper is in overall command and is responsible for the general actions of the team, as well as the details of setting up and taking a shot, the observer has a number of specific duties of his own. On the approach to the target area, he will be responsible for navigating the team into general proximity and using any necessary equipment such as communications gear or navigational aids. During the stalk, i.e. the approach to the shooting

Sniper Team

In a three-man team, each man has a clear responsibility. The sniper is the primary shooter and the team leader. The observer spots his shots and may take over some of the shooting in a protracted engagement. Meanwhile the third man provides security and all-round observation.

Spotting the Fall of a Shot

A sniper's field of view through his telescopic sight is rather limited. His observer has a wider field of view, which is not disrupted by the recoil of the weapon. Even if the sniper loses sight of his target and cannot see

where a given shot went, his observer will be able to inform him. This allows corrections to be made if necessary, or the effects of a shot to be properly logged.

position, the sniper leads and the observer provides security. This includes both watching for threats and defending the team at need, as well as more passive measures such as obliterating tracks. The observer is also responsible for cleaning up after the team by dismantling the hide and ensuring that no evidence is left behind.

Once in position the observer's primary role is to assist and support the sniper. He is still concerned with security of course, and will also record the team's actions and any observations made. He searches for

Recording Data

Modern spotting telescopes are digital camera-compatible, enabling the sniper team to record the effects of a shot or to pass accurate reconnaissance data back to base either upon their return or, if necessary, by transmitting the images.

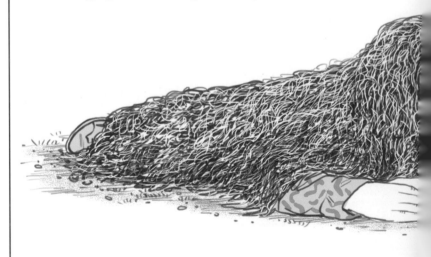

targets and marks the shot, i.e. he watches to see where the sniper's bullet goes. If a hit is made, the observer tells the sniper so. If not, the observer provides information on the fall of shot, enabling the sniper to correct his aim. Some duties are shared, no matter whose official responsibility they may be. Estimates of wind, range and other environmental conditions are a team effort, although the observer will normally use any devices such as laser rangefinders. He will also be responsible for drawing range cards.

Sniper with Anti-Materiel Rifle

Anti-materiel rifles are too bulky for most sniping applications, and their immense power is wasted in many situations. However, in an urban environment they can be useful in penetrating cover or stopping vehicles.

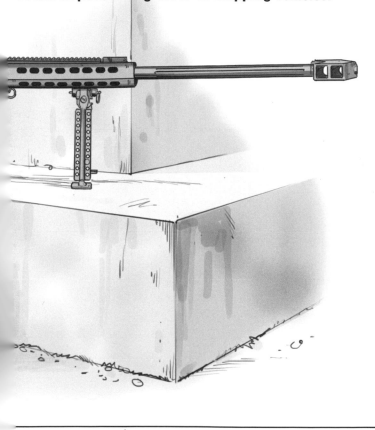

Hostage Situation

If possible, it is best to negotiate a peaceful hand-over of hostages, but if the hostage-takers seem likely to harm or kill their victims, the only effective response may be a head shot. If the nervous system is disabled, the gunman will not be able to pull the trigger and his hostage, though traumatised, will survive.

Specialist Applications

Snipers normally shoot at personnel targets from a static position. However, there are other applications of snipers' skills and weaponry, each with its own challenges.

Precision shooting with powerful weapons is one way to disable a suspect vehicle. A round from a large-calibre anti-materiel rifle striking the engine block of most cars or boats will cause catastrophic damage. A stopped engine halts the vehicle and allows arrests to be made, but shooting at a moving vehicle is a difficult challenge at the best of times. On land, the sniper must be aware of his surroundings at all times, as there are likely to be innocent people in the vicinity.

A large round may go right through a vehicle if it does not strike the engine block directly, and may scatter pieces of the engine or vehicle body in all directions. The round and/or larger pieces of the engine may ricochet for some distance. There is also the risk of injuring the suspects, which can raise additional problems. Law-enforcement officers are permitted to use lethal force to protect themselves or others from danger, but in cases where suspects are fleeing but not firing on the officers, the situation is less clear-cut.

Complicated Situations

If the fleeing suspects are considered extremely dangerous, then firing at them may be justified, and shooting to disable a vehicle, either by firing at the engine or tyres, is often justifiable. If the suspects are injured as a result of the shot, either directly or by secondary projectiles, then the legal situation becomes complex. A clean hit that takes out the engine without harming the suspects solves this problem.

Fast-moving vehicles

Even just hitting a fast-moving vehicle is a challenge; achieving a solid impact on the engine block is more difficult still. Snipers operating against fast drugs-smuggling boats have an even harder time of it. A sniper may be called upon to shoot from a moving helicopter at the engines of a boat skimming across the waves at speed and not moving in a straight line. Under most circumstances, refusing to stop is not grounds for the use of lethal force, so the sniper must make a very precise shot under extremely trying conditions.

Hostage situations

Snipers attempting to contain a hostage situation or armed stand-off must also be extremely careful to avoid collateral casualties. There are many people who will ask why an ordinary police officer could not have 'just shot the gun out of his hand' after an incident, and who expect miracles from experts such as snipers or hostage-rescue teams.

In many cases the general public are unwilling to accept that a messy outcome was the best that could be achieved under the circumstances, and an agency that manages to save the majority of hostages from otherwise certain death may be rewarded with lawsuits because an impossible expectation was not achieved. Thus, any employment of force must be as 'clean' as possible. Shots must be taken with extreme precision and with thought to where the bullet will go afterwards. A round that takes out a hostage-taker may pose a hazard to innocents downrange or ricochet dangerously, even after passing through the target.

A sniper in a hostage situation may have to make a no-shoot decision even while being ordered to fire. He is the one who bears the responsibility for his shot, and it is he who can best see whether circumstances are right. He may also have to shoot through glass or

M-87R Heavy Sniper Rifle

Sniping and target rifles have much in common. Many feature ergonomic stocks and grips. Flash hiders, muzzle brakes and integral bipods are more commonly a feature of military weapons, however; few target rifles have them.

a light wall, which will make targeting problematical. In a hostage-rescue scenario, the sniper will normally need to disable his target instantly. Whether or not the target dies is academic, although most shots that cause instant disablement also kill. What matters is that the target must not have time even to pull a trigger if innocent lives are to be saved.

A military sniper can often settle for a hit that will likely kill the target from blood loss or shock, although

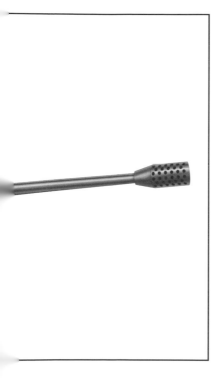

this is not an ideal situation. If the goal is to eliminate an enemy officer and that officer dies two hours after being shot, then the job is done. A hostage-taker who is not dropped instantly may kill several hostages before collapsing, and this is not acceptable.

Anti-materiel shooting

Extreme precision is also necessary when undertaking anti-materiel shooting. Targets such as support weapons and antennae require a clean hit at a vital point, which is often very small. Similarly, Explosive Ordnance Disposal (EOD) work requires considerable precision. This is not normally a sniper's task, but heavy anti-materiel rifles are sometimes used to disable or destroy explosive devices.

A static object may not be a very hard target, but there is a limit to how close the shooter can get before being endangered by the explosion likely to result from a successful shot. Therefore, the ability to keep a safe distance and destroy the device, unexploded shell or suspect object with an explosive rifle bullet is a valuable tool. Improvised Explosive Devices (IEDs) are increasingly used to attack security forces, and shooting is a method of disposal that allows them to be tackled without deploying hugely expensive equipment or placing expert personnel at risk.

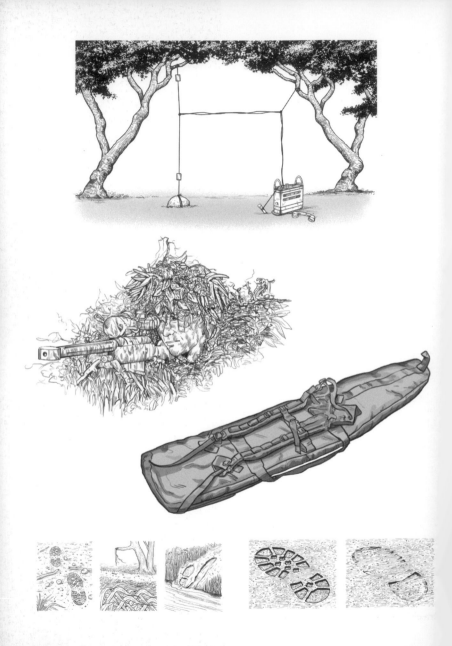

From the moment a sniper leaves the safety of his base, he is in danger from the enemy and also, to some extent, from his own side. Friendlies who do not realize that a sniper is present may call in artillery fire or air support on the area where the sniper is concealed. They may 'reconnoitre by fire' a possible enemy hiding place or respond to a shot without realizing that one of their own is on the receiving end. A sniper being killed or injured by indiscriminate or even friendly fire is as ironic as it is undesirable.

Capture

To some extent, therefore, the sniper may be in danger from both sides. However, the threat from hostiles is much greater and does not always end if the sniper is captured. In the hands of enemies willing to execute or maltreat prisoners, snipers are probably no worse off than anyone else, but even otherwise honourable troops may have sufficient hatred for snipers to beat or even shoot them out of hand after their surrender.

It is by no means certain that a captured sniper will be ill treated, but

. . . .
Snipers must be resourceful in day-to-day survival, and doubly so when attempting to bypass enemy security positions or to evade pursuit. They must develop some tracking and evading skills, too.

8

Snipers spend a long time in the field, often lightly equipped. They must be resourceful in day-to-day survival.

The Sniper in the Field

combat troops do have a special hatred for enemy snipers, especially if they have recently taken casualties from sniping. A sniper who thinks he may be forced to surrender, or who is part of a force that has been ordered to do so, may be able to take steps to look like an ordinary infantryman.

Unless there is an unusual degree of malice between combatants, many combat troops tend to see the enemy as 'just guys like us' and will be humane towards any who are no longer a threat.

A sniper who can divest himself of his camouflage and any

Expedient Patrol Antennae

Snipers must be adept in operating a variety of communications equipment, and at making it work in the field. An expedient antenna can be rigged using any handy support; since a sniper will usually try to stay away from built-up areas this is likely to be a tree.

specialist equipment or insignia, and who is caught with an ordinary rifle rather than a sniping weapon, will be treated no better or worse than any other infantryman. Alternatively, he may decide that his best hope is to avoid capture. Extreme risks become more acceptable when the alternative is surrendering to an enemy known to shoot snipers out of hand. At times in history, snipers have simply held their ground and kept shooting until overrun if there was no way out, meeting death on their own terms and perhaps buying time for a miracle.

Mission Planning

No sniper wants ever to be put in the position of having to choose between surrendering to an enemy who may well murder him, fighting to the last or making a desperate bid to escape a disastrous situation. Ideally, the sniper carries out his mission and returns to base without ever being spotted by the enemy. This process is greatly aided by effective mission planning and good intelligence, the latter provided by other snipers acting as scouts, as part of an ongoing process. Every sniper's observations add to the body of knowledge available to friendly forces and make it easier to plan a successful mission that will not unduly endanger the sniper team.

Mission planning makes use of whatever information is available, ranging from maps and photographs of the target area to intelligence estimates of enemy strength and dispositions. The level of enemy and civilian activity in the area is important, as is the quality of enemy troops deployed, if known. While this information cannot be totally relied upon, it will give an indication of the competence and speed of any response, and the likelihood of effective patrols or other security measures.

Infiltrate and Exfiltrate

Similarly, a sniper team cannot rely 100 per cent on estimates of enemy positions and patrol routes but can make some useful inferences. There is always the risk that a particularly diligent enemy patrol will search the most difficult terrain on their route, but it is more likely that patrols will follow the obvious routes, especially if the sector has seen little or no action. There will be some points that are certain to be watched or garrisoned, and others where there is little point in maintaining a defensive position. This allows the sniper team to plan a route around the areas that must definitely be avoided and offers the best chance to infiltrate and exfiltrate the target area without coming into contact with hostiles.

The mission plan also needs to include information on friendlies in the area and communications protocols such as call signs. The degree of support and resupply available to the team, and the form these will take, need to be clearly outlined. A last-minute update may be available, but all data necessary to planning the mission in general terms must be known in advance. A major change in the situation may necessitate the mission being replanned or scrubbed; attempting to proceed and update the plan on the fly is a risky business at best.

Desperate Measures

There are occasions when a proper plan is not available and there is no

Helicopter Drop

A helicopter that flies to a particular point, stops for a while and then goes home is a rather obvious clue that something is going on. If a sniper team is inserted or resupplied by helicopter, the aircraft will normally 'visit' several points around the general area to at least create some confusion about exactly where the drop-off occurred.

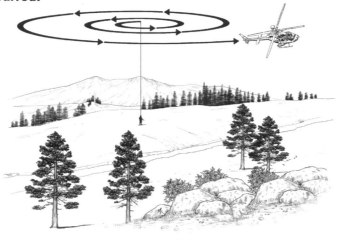

time to create one. Under such circumstances, permission to proceed with whatever desperate enterprise the sniper team has in mind would normally be refused. However, there have been occasions where snipers have insisted upon going to the assistance of comrades who were in mortal danger. This occurred in Mogadishu in 1993, when two snipers were inserted into the combat zone at their own insistence to protect downed helicopter pilots.

With no prospect of immediate support and no plan beyond defending the wreck, the snipers were able to hold off the enemy for a time before being killed. They knew the risks when they requested permission to deploy (for the third time; they were turned down twice before being allowed to go) but chose to try

A Good Plan

Although it takes at least as much courage to crawl for hours past enemy sentries as to enter a firefight, a good plan does limit the risks somewhat. A good plan is not merely a blueprint for how the mission would go if everything played out perfectly; it is a framework for co-operation and a basis from which a new plan can be created when circumstances change. Time spent planning a mission can be viewed as a form of life insurance for the sniper team.

A good plan not only assists the sniper in the execution of his mission, but also enables him to make sound decisions once deployed. Few plans survive intact once the enemy comes into proximity, but a good planning session allows the sniper to work through possibilities and to have at least an idea of what might change. He will also know, in general terms at least, what is expected of him and any other forces involved. He can, therefore, make a best guess about what friendlies might do when the plan comes unstuck, and put himself in a position to support them or benefit from their assistance.

anyway. This sort of seat-of-the-pants heroism is the opposite of what snipers plan for, but in this case there was no alternative. Both were posthumously awarded the Congressional Medal of Honor.

Rules and Risks

Any mission plan must take note of prevailing rules of engagement and collateral risks. In a clearly defined war zone, the sniper team is at liberty to engage any identified enemy target, but there are circumstances where the snipers' task is made even harder by limits placed on who can be fired upon. The rules may be as simple as a requirement to identify a weapon before assuming that a potential target is a hostile and therefore a legitimate target, or they may be complex and situational. In a law-enforcement context, the rules of engagement tend to be very strict and are based upon actions as well as armament. A gunman who is not apparently going to make a hostile move in the imminent future may be a perfect target but be off limits.

Snipers on the Move

A sniper team must balance the needs of mobility with concealment. On the one hand, if the team is detected, then the mission may be compromised, but a team that never reaches the mission objective is equally useless. Movement tends to be quicker in safer areas, slowing down as the sniper nears his target.

One way to get a sniper into position quickly is to deploy directly from a helicopter or vehicle. This can be somewhat obvious if the enemy is observing, but if there is a lot of activity, then the enemy may not be able to tell where the sniper has been landed. Alternatively, a patrol or strike force can be landed by helicopter and then withdrawn after its mission is complete, leaving behind the sniper team in a concealed position.

Vehicle Convoys

A vehicle convoy can also drop off the sniper team as it proceeds on its own mission. This can be done to get the team close to an objective or may be a security precaution for the convoy itself. Infantry patrols are used to conceal relatively rapid sniper movement in the same way; the patrol moves through its designated zone and stops at various points along the way. The sniper team moves with the patrol until a suitable moment to slip away presents itself. The patrol can act as a diversion, drawing the attention of enemy observers while

Shoot and Move

A sniper who stays in the same place invites discovery and/or return fire. A policy of shooting then moving to another position is generally wise, although there are always risks inherent in moving when the enemy is in proximity. Thus a sniper will tend to pick his shots carefully to get the best effect from them. The number he can fire before his position becomes unviable is fairly limited; he must balance the risk against the potential gain.

Signs of Movement

Snipers, and those searching for them, must be adept at noticing the signs of human passage. Conversely, snipers are trained to leave behind minimal traces and to either eliminate those they do create or to leave a false trail when necessary.

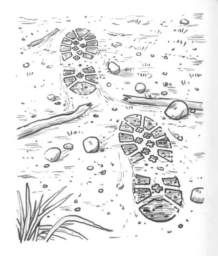

Broken sticks

Fresh footprint

Disturbed foliage

Slip marks/footprints on a river bank

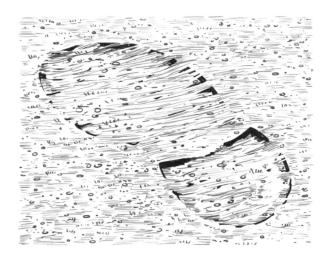

Footprint affected by rainfall

the snipers move into their concealed position.

The same method can be used to cover the pick-up of a sniper team. If what appears to be a routine patrol comes back with a couple of extra soldiers, this may well escape enemy notice. This helps preserve the secrecy of the snipers' hide and may conceal the fact that a sniper team was ever deployed in the area. A team that has been observing or did not get a suitable opportunity to shoot may want to return to the same area later;

Walking Backwards

One way to leave a false trail is to walk past some concealing object for a distance, then go back to it walking backwards in your own boot prints. Anyone following the main trail may miss the point where the real path diverges.

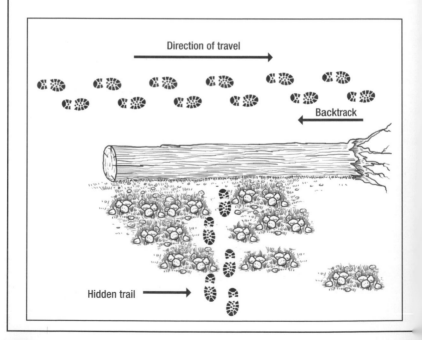

Direction of travel

Backtrack

Hidden trail →

if the enemy never knew they were there, then a search is unlikely.

Camouflage

If a sniper team is moving with other infantry, it is important that they are not identifiable as snipers. Specialist equipment such as ghillie suits and spotting scopes are kept out of sight, and rifles are held close to the body where it will be hard for an observer to distinguish their unique nature. The snipers will wear the same uniform and carry the same

Once the false trail ends, however, a search may find the real one. To get around this, it is possible to 'end' the false trail at hard ground where few prints would be left, hoping that enemy trackers will try to pick up the trail at the far side rather than doubling back.

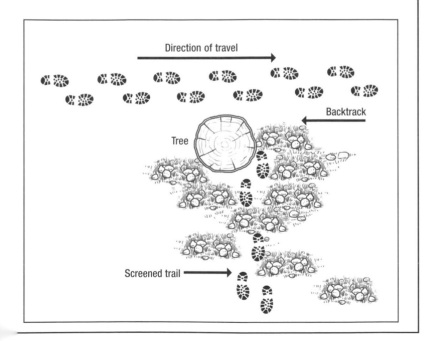

Direction of travel

Backtrack

Tree

Screened trail →

equipment as the other infantrymen, and take care not to stand out in any way. The intent is that they look like they are a part of the infantry force while they are with it. Once company has been parted, the snipers can don full camouflage and stash any equipment that is not

Field Expedient Camouflage

A ghillie suit can be improvised in the field from foliage, sacking and any cloth of a suitably dull colour that can be found. Even where a carefully-made ghillie suit is available, it will often be enhanced with bits and pieces of local foliage. Nothing blends in better than what was already there.

necessary for their mission. Normally, the snipers will remain static while the rest of the force moves off, allowing the snipers to watch for a while and see if they or the infantry force have been observed. The same applies when a sniper team has been conveyed to a

target area by an infantry force acting as a security element. The security element moves off to its next destination and the sniper team waits a while before beginning its approach to the target area.

When moving independently of other troops, a sniper team can travel surprisingly quickly overland without compromising stealth. There

Cutting the Corner

A false trail can be left when joining a path or road, suggesting that the sniper team has moved in one direction when they are in fact headed elsewhere. If the track has a hard surface, a few direction-indicating prints can be 'accidentally' left as the team joins the track, doubling back on the hard surface to continue on the other side.

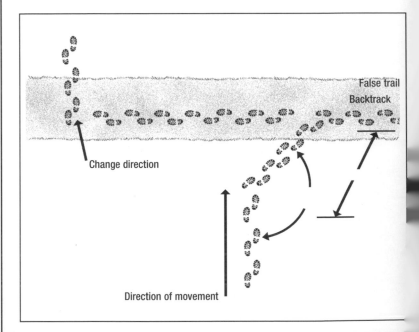

False trail
Backtrack
Change direction
Direction of movement

are no absolutes of course, but snipers are taught how to balance stealth with speed, moving at an acceptable pace without becoming easy to detect. Their habitual observation of their surroundings makes it likely that the sniper team will become aware of hostiles in the area before they are spotted, and can assume a more stealthy posture.

Fish Hook

The 'fish hook' can be used to ambush suspected trackers, or to observe to see if anyone is following. As the name suggests, the sniper team doubles back in a wide arc to a position of concealment, from which they can see their previous route.

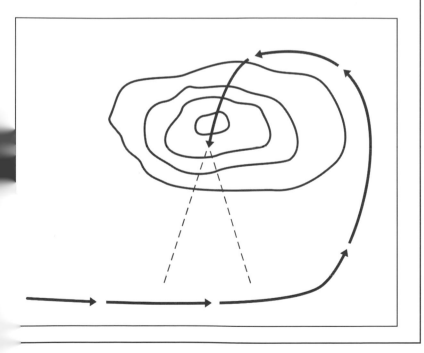

Oblique Movement

The shortest and most direct route is rarely the best for snipers in the field. Each obstacle or hazard must be painstakingly circumvented, and always with a view to concealment. A sniper team that becomes impatient or lazy will likely be discovered and, at best, captured.

Thinking Differently

A sniper has to unlearn some habits that most people develop. An ability to pick out the easiest route is beneficial to most people who have to move overland, but a sniper needs to avoid taking the obvious path. Not only is this where an ambush or booby-trap is most likely, but enemy patrols will also move by the easiest route, making an encounter more likely.

Thus, a sniper needs to be able to pick a route that will not slow him down or tire him too much, but which also offers good concealment and is unlikely to be used by hostiles. This means that the sniper should stay off paths and roads wherever possible, and needs to be very careful when crossing them. On the other hand, dense foliage may offer good visual concealment, although moving through it may be either excessively slow or noisy.

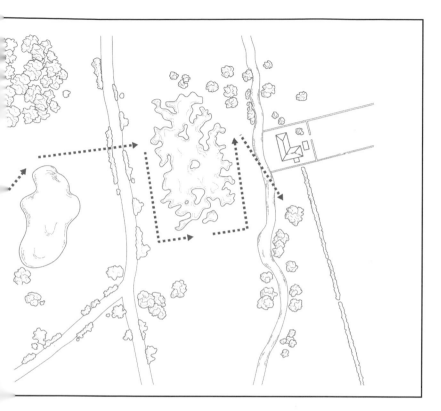

When moving during daylight, the main threat is visual detection. The human eye is drawn to movement, especially if it is sudden or jerky. Steady and fairly slow movement is less likely to attract attention. Shadows must be exploited to their fullest extent and silhouettes are a giveaway that must be avoided. Snipers know better than to 'skyline' themselves by moving along the top of a ridge, and should be mindful of what is behind them. Their camouflage is less effective when they are outlined against a building wall.

Sight, Sound and Smell

Poor camouflage can also give away the sniper's presence. The eye resolves familiar outlines quickly, so camouflage should cause the sniper to blend softly into the background rather than having a hard outline. The outlines of helmets and rifles in

Moving Predictably

Inexperienced or badly trained shooters may make themselves more vulnerable by predictable movement. A marksman who always takes one shot then moves to a different window in the same building, or one who shoots from one end of a wall then crawls to the other for his next shot, may fool infantrymen who are engaged in a close-range fight with his comrades and who do not have time to analyze his pattern. A watching sniper will have little trouble in predicting what the gunman will do next, however, and may even have a shot already lined up when he gets into firing position. Likewise, a shooter who always uses the same position, then vacates it, may be invulnerable to fire when he is in position, but be an easy target for a sniper as he exits the back door or across the rooftops after his shot.

Drag Bag

A 'drag bag' is designed to carry the rifle and other equipment in the field, and to protect it while being dragged through the most rugged terrain.

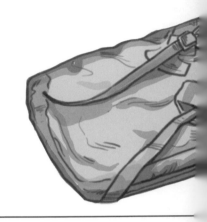

particular need to be 'broken up' or softened. A rifle may be carried in a 'drag bag', which is camouflaged similarly to a ghillie suit. Reflections from lenses or any shiny surface must be avoided; this goes beyond poor camouflage – light glinting from binocular lenses actively advertises the sniper's presence.

At night, the sniper team is less likely to be spotted but it may well be heard or even smelled. Equipment is secured so that it does not rattle wher

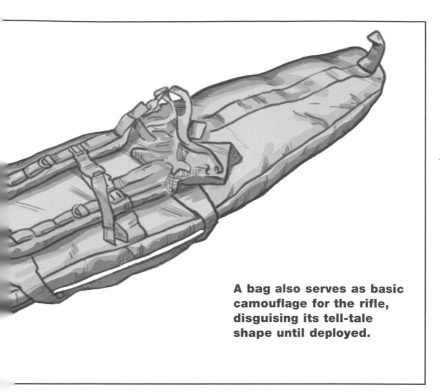

A bag also serves as basic camouflage for the rifle, disguising its tell-tale shape until deployed.

the sniper moves, and it is important not to forget to re-secure anything that is used. While an occasional sound may not be recognized or even heard, voices will be quickly recognized as such by patrols in the area. Sounds can be heard over a considerable distance at night, and scent can also carry a long way. Snipers avoid strongly scented shaving products, soaps and the like when in the field, and must be careful about cooking smells as well.

A sniper team on the move works on the assumption that the enemy has patrols and sentries in the area, and that they are alert. Movement is slow and deliberate and is planned in advance. The team will stop and observe before moving to the next position, and will be aware not only of their own actions, but also of secondary indicators such as the movement of undergrowth as they pass through, and the possibility of disturbing birds or wildlife.

As the sniper team gets close to the objective, movement must become ever slower and stealthier. If the team is approached by an enemy patrol, it will freeze and wait for the patrol to pass by. The sniper team would only initiate combat if discovery were imminent.

In this case, the intent is to break contact and escape as quickly as possible. Discovery in this manner will often require the team to scrub the mission and return to base; an alerted enemy makes further contacts likely and success much harder to achieve.

If the sniper team is able to avoid contact and get close to the target, the final move into shooting position will be slow and careful. If a shot is to be made from a significant distance, or if the sniper is to wait for a target to present itself, then a hide may be constructed.

This is not always possible; a sniper will often have to move into a naturally concealed position to make his shot.

. . . .

As a general rule the closer a sniper gets to the enemy, the slower he moves. The first kilometres of an approach might be covered at some speed. Progress is measured in metres or even centimetres when close to the enemy.

9

Snipers need to develop patience and the ability to move without being detected in order to be effective.

In Proximity to the Enemy

Low and Medium Crawl

The sniper's low crawl (left) is incredibly slow. An observer may not even notice the sniper moving. Immense patience is needed when moving into position in this manner, as well as the courage and faith to carry out the crawl even while there are hostiles in the area. A medium crawl (right) is used when it is reasonably safe to move a little faster.

Legs together
– push with toes

Pull with fingers
– hold weapon by sling

Lie flat on the ground
– push with legs

Pull with arms
– hold weapon by sling

'Bullet Magnets'

A sniper's task becomes more difficult if he is under fire, even if that fire is actually directed at the ham-fisted mob of gunmen who are blasting away from a nearby position. A sniper who sets up too close to friendlies may become collateral damage if they draw artillery fire or if their position is assaulted by the enemy. However, a sniper can actually benefit from friendlies who draw attention in this manner, as long as he is sufficiently distant that hostiles who are engaging his comrades are not looking in his direction.

Crawling

If there is natural cover such as 'dead ground' (i.e. low-lying ground screened from the enemy's vision by intervening terrain), then the sniper may be able, more or less, to walk into his position. More likely, he will have to crawl. The lower a sniper's body is to the ground and the slower his movements, the less likely he is to be spotted. However, a low crawl is very slow, so a sniper has different modes of crawling movement available for differing situations.

Hands and knees crawl

The fastest form of crawling movement is also the least stealthy. The sniper cradles his rifle under one arm, with the scope in his armpit. He supports himself with his other hand and on his knees, and moves as quickly as necessary into position. His body is completely clear of the ground.

High crawl

The high crawl is similar to that used by infantry. The rifle is cradled across the sniper's body in the crook of his arms. He supports himself on elbows and knees, with his body just off the ground.

Medium crawl

The medium crawl is used as a compromise between speed and stealth. The rifle is held and moved using its sling. The sniper is flat to the ground and alternately pushes himself with his legs and pulls with his arms.

Low crawl

The low crawl is used when extreme stealth is required. Again, the sniper pulls the rifle by its sling. The sniper's legs are together. He pushes himself with his toes and

Low Crawl

When crawling, the rifle can be dragged by the sling or cradled in the arms. This has the advantage that it is partially concealed by the ghillie suit, but the barrel and stock do inevitably move about when crawling, and may attract attention.

Support body with
elbows and knees

Weapon cradled in
arms

High Crawl

When it is safe to crawl on hands and knees, a sniper can move somewhat faster. This is relative of course; crawling is never the quickest of ways to get around.

Scope cradled in armpit

Body supported by knees and hands – weapon carried under arm

pulls with his fingers. This form of movement is very slow and is generally used only when the enemy is in close proximity or while moving into a firing position.

Attack or Threat
In the event where a sniper team

comes under attack or is threatened by the approach of enemy forces, the preferred solution is to hide or displace to avoid contact. All that may be necessary in the case of a patrol may be to halt and wait for a while, but sometimes a sniper team is forced to move out of the path of

get out of the target area as quickly as possible without compromising concealment. Once clear of the target area, the team will seek an alternative route to the target. In the case of air attack, the best defence is concealment. Aircraft cannot loiter long and tend to move fast, so once a target is lost, it is unlikely to be found again. The sniper team will therefore find the best positions available, moving to better cover as soon as the aircraft is out of sight. If it comes back for another pass, the team remains still and hidden until it departs.

If contact with a hostile patrol is made, the team needs to break contact quickly. Smoke grenades obscure the team's retreat while heavy fire from the observer's weapon may eliminate some hostiles and encourage others to take cover and lose sight of the sniper team. Immediate pursuit can be discouraged by halting to shoot – accurate fire on the first to move in pursuit may discourage the rest of the hostiles, but the aim always is to break contact, using movement and concealment. Once the hostiles lose sight of the sniper team, they may be unable to regain contact, even if the snipers are hiding close nearby.

hostiles, or must find an alternative firing position because a patrol has established itself nearby.

Mortar fire

In the event that a sniper team comes under indirect fire from mortars or artillery, the team needs to

Dogs

Dogs are a serious threat to a sniper team. Fortunately, they are not usually deployed with routine patrols.

If dogs are present, they will be able to track a sniper by scent long after he has passed. Guard dogs in static locations are a similar hazard. Shooting the dogs presents a solution of sorts, and may be the preferred option when penetrating a defended base for a strike. The silenced pistol favoured by Vietnam War-era US special forces personnel became nicknamed the 'hush puppy' for this reason.

However, shooting dogs is not an ideal solution – apart from anything else, there are usually several of them and their handlers are accompanied by troops who will readily shoot back. It is better, where possible, to break contact by using running water to mask the sniper's scent. Simply crossing the water is not enough – dogs can pick up the scent easily enough on the far bank. Instead, the sniper team needs to move upstream or down, perhaps leaving and entering the water several times to confuse their trail.

In any case, a sniper team that is compromised must rendezvous with friendly troops, return to base or call for extraction. If a security element is located nearby, the two groups will have an agreed rendezvous point and will head for it.

However, a sniper team operating deep in enemy territory may have to travel some distance to reach safety. The temptation to rush must be ignored; if the team does not have

hostiles in close pursuit, then its exfiltration will be every bit as stealthy as its infiltration.

Return to Safety

Maintaining stealth and concealment while in proximity to the enemy is difficult enough. Doing so throughout a long mission is even harder, yet this is precisely what a sniper team must do. One of the most dangerous moments in a mission occurs after an enemy patrol has passed by or the team has moved out of the firing position after taking a shot or making observations. The mission is not over at this point, even if the objective has been accomplished. The team must now return to safety without becoming sloppy and inviting discovery.

This requires considerable self-discipline. It is tempting to relax after escaping a dangerous situation, but for a sniper team this is highly dangerous. The ancient samurai used to have a saying: 'After the battle is over, tighten your helmet straps.' This applies equally well to a sniper team that has completed its mission or broken contact with pursuers. 'Safe' means back at base, debriefing completed and maybe a hot meal on the table. 'Safe' is not defined by no longer being in close contact with the enemy; snipers that remember this live much longer.

Walking

If there is little chance of observation, or where concealment is good, the sniper may decide to walk instead of crawling. He will still move slowly and cautiously, and will crouch or bend to alter the distinctive shape of a man walking upright.

Crouch with body bent forward and knees bent

Muzzle pointed down

10

There are times when a sniper has to shoot from wherever he may be, for example during a contact on the move or when shooting from a vehicle. However, if at all possible, he will select, improve or construct a position. A good position offers both concealment and support for the weapon when shooting, but there is more to selecting a position than this.

A suitable position has a good field of fire and observation, and is at least 300m (328 yards) from the target with some kind of obstruction between sniper and target. This makes a quick reaction difficult even if the sniper's position is detected. Routes into and out of the position are also important. They should offer concealment from observation and cover from fire. Ideally, this will enable the sniper to slip away after taking his shot, but if he is spotted, at least he will be protected from enemy fire as he retires.

During World War II, church towers became notorious as sniper haunts. A tower does offer some of the advantages listed above. Field of fire is good and the elevated position allows long-range observation of an approaching enemy force. The

· ·

The nature of a sniper's position and the amount of effort put into preparing it depends greatly upon the availability of natural cover and how much time is available.

The Sniper in Position

Tentative Position Areas

The sniper team approaches its tentative final firing position (TFFP) from a position of concealment. Once in the area, a final firing position is selected, based on its merits for concealment, visibility of the target area and ease of withdrawal.

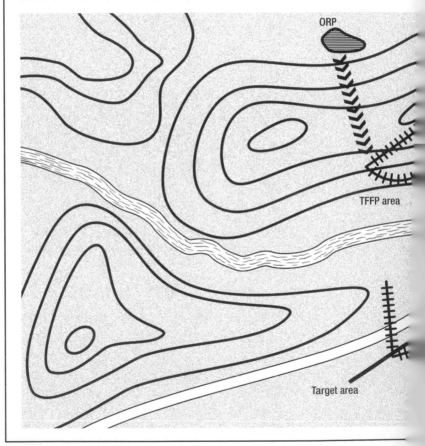

ORP

TFFP area

Target area

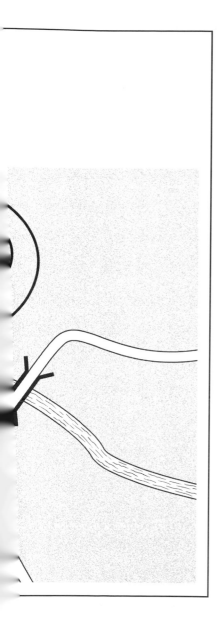

structure of the tower offered some cover from small-arms fire, and it was unlikely that hostiles would be able to sneak up on the sniper. However, towers were also obvious places to shoot from and would attract counter-fire from anti-tank guns and similar heavy weapons against which they offered no protection. Using a tower to shoot from could also trap the sniper even if he was not killed by return fire. Escape had to be well timed; a church tower is not a place that can be left quickly.

Important Factors

There are several factors that make an otherwise excellent position unusable. The lack of a suitable escape route is one, and any position that is obviously a good place for a sniper should be avoided. An elevated spot looking along a road, or an isolated high point, will be too obvious; the sniper should set up elsewhere, as these spots will attract scrutiny from alert hostiles even if there is no shot, and will likely be raked by speculative return fire as soon as the sniper goes to work. Instead, the sniper must find a location that balances suitability as a firing point against the likelihood of detection.

A sniper team does not move directly into position, but instead halts at a previously selected Objective Rally Point (ORP). From here, the firing position is observed or, if one has not already been chosen, the team looks for somewhere suitable. If the firing position appears to be safe and suitable, one member of the team moves cautiously up and enters while the other remains at the ORP to give cover. Once the first team member is in position, the second moves up.

Once the team is sure that the position is suitable and has not been compromised, preparations can begin for the shot. If a hide is to be built, construction begins. The team will also prepare and lay out equipment and draw up range cards. Depending on the mission, the team may be in this position for some time. If so, a fairly elaborate hide is a possibility, but for a fairly short stay the team will rely mainly on concealment already in place.

Hides in an Urban Area

In an urban area, the team will probably set up on a rooftop or a room in an upper floor. An urban position can be improved by breaking holes in a wall or removing roof tiles to allow a sniper team located in an attic space to see and shoot out. Rifle barrels poking out through the brickwork are something of a giveaway, so the team must be careful to stay back from the hole.

Likewise, most light coming in through a window or hole in a wall

Loopholes

Loopholes in Hide Positions

The last thing a sniper wants to do is poke his rifle through his carefully constructed concealment, possibly giving away his position. Loopholes must be created when the hide is constructed, and sited to give a clear shot at the target area.

illuminates the nearest part of the room; the back of a room is much more poorly illuminated. A sniper team will set up at the back of a room, concealed by shadows, rather than in the light. This does limit their field of view but it makes them very hard to locate. Furniture can be moved to improve the firing position. For example, a table moved into the

Urban Sniper Hide

Urban areas offer the sniper team plenty to work with when setting up a hide. Furniture provides a rest at a comfortable and convenient height, while the curtain is secured to prevent outside observers from seeing into the hide.

room allows the sniper to lie prone but still see out of the window.

A sniper team could also set up away from buildings in some circumstances. This might limit their line of sight, however. Suitable locations include any cluttered environment such as rubble or wrecked cars. A location in the open has the additional drawback that it

Crawlspace

The crawlspace between floors of a tall building makes an excellent hide. A small loophole is far less likely to give away the sniper's position than if he sets up in an office surrounded by a vast expanse of glass or broken windows.

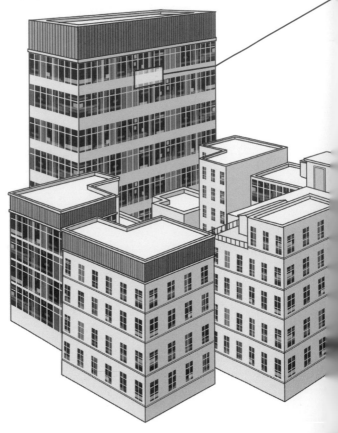

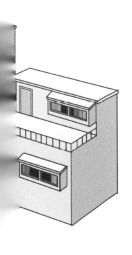

can be difficult to enter and withdraw from without being seen, but there may be few alternatives.

An urban position can be strengthened by sandbags or rubble, without necessarily being detected. Sandbags that cannot be seen inside a building will not be detected, while rubble may not be out of place in an urban combat zone. Obviously, any work that is done needs to be carried out in secrecy, but this is possible if there is a lot of friendly activity in the area – work on the sniper position may be lost amid the more general movements of troops.

Noncombatant Risk

The main risk when setting up a hide in urban terrain is observation by hostiles or discovery by noncombatants who may give away the snipers' position by their actions. Even if noncombatants do not deliberately inform the enemy of what they have seen, they may be questioned or may simply react in a manner that suggests snipers are present. The sniper team has no way of knowing whether the people who have spotted them will reveal their location or not, and may have to abandon a position just in case.

Hides in a Rural Area

In a rural area there may be sufficient cover to rely on personal camouflage. This is necessary when a firing position must quickly be found. A

Keep Your Distance

An effective sniper, whether formally trained or simply a gunman who has learned his trade in the field, will take care to conceal his position and to avoid drawing attention to himself when he is not shooting. This good work can be undone by outside factors, including inquisitive wildlife and less stealthy comrades. Operating at a suitable distance from such hazards makes the sniper's task much easier.

depression in the ground, thick foliage, rocks or fallen trees can all offer a workable hasty position, which can then be improved with whatever natural camouflage is quickly available.

A position of this sort may not be ideally located and may have blind spots, but it does offer the advantage that it can be occupied without lengthy preparation. It will still be necessary to create loopholes to shoot through, and to ensure that local foliage is not disturbed by the weapon's muzzle flash. Therefore, some measure of preparation will be advantageous.

If there is a little time, a shallow scrape can be created in the ground, ideally behind naturally occurring cover. This allows the sniper to disappear completely until he raises his head and rifle. Earth from the scrape can be used to fill sandbags for protection and to support the weapon. This kind of position can be built quickly and does not require extensive preparation, but there is still some digging to be done. The team must also be careful not to alert enemy patrols during the preparation phase.

Improving Concealment

If there is time to build a fairly permanent position, the sniper team can do much to improve the concealment, comfort and protection offered by their hide. A natural depression can be enlarged or a trench dug, but this creates the problem of what to do with the dirt. It must be removed from the hide site and somehow concealed; a fresh pile of earth may arouse suspicions in even the most complacent of troops.

If overhead cover is to be provided, then this must be sturdy and not prone to damage from wind or rain. Timbers covered with dirt and then

Belly Hide Position

A belly hide requires much less work to construct than one in which the sniper can stand or sit. Digging may not be necessary if a natural hollow or gap between two higher points such as logs or rocks can be utilized.

Entrance

15cm (6in)
of earth

Semi-permanent Hide Position

For long-term observation or when attempting to deny an area to the enemy, a semi-permanent hide can be constructed. Rotating teams into and out of the hide runs the risk that its location will be betrayed, so such a hide has a finite useful lifespan.

Entrance

camouflaged will work well, but obtaining suitable logs is a problem that must be approached carefully. Deadfall wood, unless it is rotten, may be a better choice than cutting branches, which will leave obvious marks. It may be possible to make use of naturally occurring cover such as a large deadfall log with a space under it, and to build this into a suitable position.

Observation

Even where the sniper's main mission is not reconnaissance, his observation skills are paramount to his success and possibly even survival. Snipers are habitually aware of their surroundings and tend to spot small things that untrained observers would not. These various skills are enhanced by the use of technological aids, but without the ability to make logical connections and inferences, their capabilities would be largely wasted.

A sniper's observation skills are a combination of keen eyesight and hearing, skilled use of equipment and a great deal of practice, plus knowledge. A sniper who understands military thinking and technique has a huge advantage over someone who must rely merely on what his eyes tell him. The sniper learns to look at the places where an object might be hidden, and into areas that a casual observer would pass over. He will notice what is not

Tent-Type OP

The tent-type observation post is made for more than one observer. Branches can be used to create the frame and a parachute used for making the inner lining. The outside should be covered in local materials so that the post blends into the surroundings.

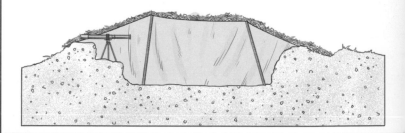

there, or what he cannot see, as readily as what is in front of him.

A skilled observer can infer the presence of a camouflaged vehicle or infantry position from as little as a vague suggestion of an outline and a shadow that is out of place. Similarly, there are only so many places where an enemy soldier or sniper can conceal himself and still be able to observe and fight. A sniper learns to spot likely hiding places and then take a closer look at them, rather than simply looking for hostiles. He also learns how not to 'lose' a suspicious object or location when he looks elsewhere, enabling him to keep several potential enemy locations under scrutiny at the same time.

Detailed observation, especially when using technological assistance such as magnifying optics or thermal-imaging equipment, is tiring

Expedient Position

When there is no time to construct a hide, the sniper team will select a position that gives good concealment and field of view, and abandon it as soon as the task is complete. This may leave no trace that the team was ever there.

Detailed Search

Once in position, the sniper team makes a detailed survey of the entire area in front of them, and estimates the range to the landmarks within it. If a target appears near one of them, its range is already known and a shot can be quickly made.

to the eyes. A sniper team swaps roles frequently, with one member observing and the other resting.

Ten minutes of observation, or at most 15, is as much as can be conducted before efficiency begins to drop off due to eye fatigue.

Quick Scan

The first phase of observation when moving into an area, or after a period during which the sniper team has not

had their surroundings under observation for whatever reason, is a hasty scan of the area. The quick scan begins close to the sniper team's position in case there is an imminent threat, then moves outwards. The observer makes a quick survey of the area using relatively low-magnification equipment such as binoculars. This rapidly familiarizes him with the surrounding area and allows him to spot anything obvious. 'Obvious' is relative, of course. What may stand out, even to a quick glance by a trained sniper, might be well hidden by the standards of ordinary soldiers. The hasty search includes a series of short looks at any locations that might be of interest, such as positions where the enemy might have concealed themselves. Unless there is some pressing reason to move away or engage an imminent threat, the sniper team then begins a more detailed search of the area.

Detailed Search

The detailed search starts close to the sniper team and is conducted with high-powered optical instruments such as an observation telescope. The entire field of view is slowly swept, concentrating on strips that overlap with the previous one and which are overlapped by the next. In the manner of a lawnmower, the observer sweeps every piece of terrain within his field of view.

Once the area has been searched, it remains under observation, with the sniper and observer alternating so that there is never a gap in observation while both avoid becoming fatigued. Hasty and detailed search techniques are used as appropriate. It is possible that the sniper team may develop a 'feeling' or a 'suspicion' about a given area, even though they cannot see anything there. This may be due to

something that seems a little out of place, or the location's suitability as a sniper hide. Such areas can be singled out for additional attention, but not to the detriment of other observation. Fixating on a suspect location can cause the sniper to miss developments elsewhere.

The observer's detailed search takes as long as it takes. It cannot be rushed: a complex area takes a long time to search, but it must be done. Hurried observations can be misinterpreted or critical details overlooked. During the observation phase, the sniper team also determines the range to various points in the field of fire. Artificial aids such as laser rangefinders can be used, but snipers are also trained to make an estimate by eye.

Tried and Tested

If artificial aids are not available, the sniper can fall back on several tried and tested methods to find range to an object. One is to use a map, usually with a strip of paper. By measuring the distance between the sniper's position and the object, the distance can be read off the map scale. The paper strip serves to carry the measurement from the map to its scale. It is marked with the sniper and target locations.

Ranges can also be estimated with good accuracy by learning to visualize a 100m (109-yard) distance. The sniper then mentally lays as

Range Estimation – Paper Strip Method

The paper strip method is very simple. The distance between two points on the map is marked on a slip of paper, which is then placed on the map's scale. The distance can then be read off the scale with great accuracy.

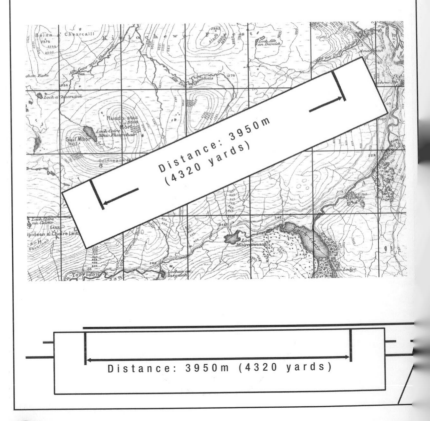

Distance: 3950m (4320 yards)

Distance: 3950m (4320 yards)

many of these 100m increments as necessary on the terrain he is observing. This works best in fairly flat, open terrain. Alternatively, he can look for an object of known size near the target, and base an estimate on his knowledge of what such an object looks like at various ranges.

The sniper's scope can be used to estimate range. If the sniper knows or can closely estimate the dimensions of an object (such as the height of a typical enemy soldier), he can use the markings on his scope to calculate range. The distance to the target in metres is given by the formula:

Known dimensions in metres x 1000
Size of target in mils on the sniper's scope

So, if an enemy soldier is estimated to be 1.8m (5.9ft) in height and

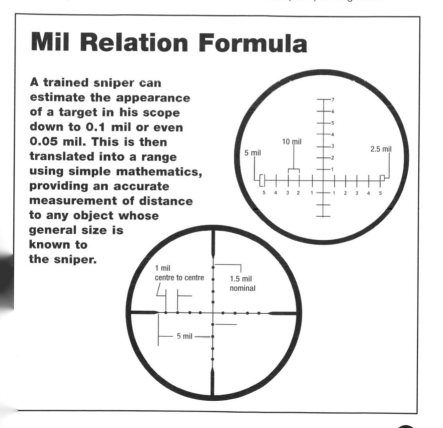

Mil Relation Formula

A trained sniper can estimate the appearance of a target in his scope down to 0.1 mil or even 0.05 mil. This is then translated into a range using simple mathematics, providing an accurate measurement of distance to any object whose general size is known to the sniper.

100 Metre Unit Measure Method

If the sniper can visualize a distance of 100m (109 yards) and mentally superimpose it on to the scene before him, he can simply count out the number of 100m increments between himself and any object he can see. Height differences can make this method difficult to use, however.

Range Card

Great artistic talent is not needed when preparing a range card. So long as the key elements of the scene are recognisably portrayed and their distance from the sniper is accurate, the card will serve its purpose.

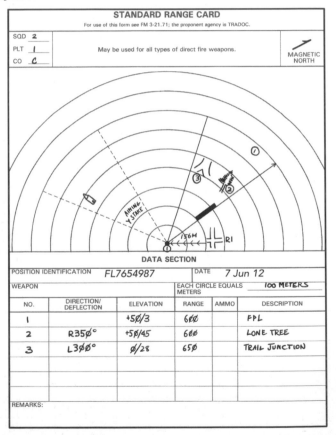

STANDARD RANGE CARD
For use of this form see FM 3-21.71; the proponent agency is TRADOC.

SQD **2**
PLT **I**
CO **C**

May be used for all types of direct fire weapons.

MAGNETIC NORTH

DATA SECTION

POSITION IDENTIFICATION	FL7654987		DATE	7 Jun 12	
WEAPON			EACH CIRCLE EQUALS METERS	**100 METERS**	

NO.	DIRECTION/ DEFLECTION	ELEVATION	RANGE	AMMO	DESCRIPTION
1		+50/3	600		FPL
2	R350°	+50/45	600		LONE TREE
3	L300°	0/28	650		TRAIL JUNCTION

REMARKS:

appears to be 3.6 mils high in the sniper's scope, then he is (1.8 x 1000)/3.6 = 500m (547 yards) away.

Of course, this method relies on an accurate estimate of the target's dimensions and works best when they are precisely known, such as the length of a specific armoured vehicle or the height of a typical urban doorway. To perform these and other calculations, a calculator and spare batteries are part of most snipers' equipment.

Landmarks

If the sniper has created a range card (this is usually done during the observation phase), he may be able simply to read off the ranges to known landmarks, as he will have already gone through the whole range-estimation process. This makes setting up a shot far faster when it becomes necessary.

However, it is rare that a target will present itself right next to a landmark at a known range. Some estimation or correction is often necessary, which can be done by bracketing the target's range between two known points. A target that is roughly one-third of the way between a 200m (218.7 yard) and a 380m (415.5 yard) point can be estimated to be about 260m (284 yards) away.

Adelbert F. Waldron (1933–45)

Waldron operated in Vietnam as part of the riverine patrol force, which meant that he often had to shoot from a moving boat. The patrol vessels were targets for all kinds of weapons, ranging from machine guns to sniper fire, and Waldron's difficult task was to counter these threats with precise return fire. He also participated in night-time operations in which snipers equipped with low-light sights pointed out targets with tracer rounds, at which point they were engaged by heavy weapons aboard the boats. He also operated on land, favouring an M21 semi-automatic rifle equipped with a suppressor and a 'starlight scope'. With this weapon he was able to engage enemy patrols at short range (for a sniper) without being located.

Flag Method

The angle between a flag and its pole gives an indication of wind speed. Dividing the angle by four gives the wind speed in miles per hour. The behaviour of the flag also indicates whether the wind is gusty or changes direction frequently.

Direction of wind

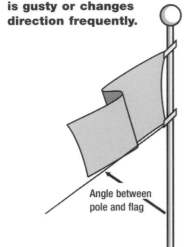

Angle between pole and flag

Terrain and Environment

Some terrain and environmental factors can influence perception of range. Very flat terrain makes objects look closer, as does looking at them from below. Targets below the sniper's position, or on the other side of low ground, can seem further away. Perspective can also confuse the eye – a relatively small object such as a human will seem more distant when located close to something much larger, for example trees or buildings. Conversely, objects that stand out against the background, or which are very easy to see, tend to appear closer than they are.

Note that although a sniper's scope can be adjusted for range, this is of little use if the range is incorrectly estimated. A scope set for the wrong range will assist the sniper in precisely putting a bullet somewhere that the target is not. This is the fault of the sniper, not the scope – it will do the job it is set up to do, regardless of whether the sniper's data is right or wrong. It is common to estimate range by more than one method to ensure an accurate measurement. An average between two or more reasonable estimates has a good chance of being correct, while wildly differing estimates suggest that the sniper needs to repeat the process with more care.

Wind Factor

Wind is often a major factor in sniping, and must be correctly estimated at various points along the path between sniper and target. Firing ranges have flags for this purpose, and it is sometimes possible to use similar objects in the field. Clothing hung out to dry, tall grass, flags, foliage and even bunting can be used to gain an estimate of wind conditions. If something so convenient is not available, the sniper has other methods at his disposal:

Nearly still air such as a breeze of under 5km/h (3mph) will make smoke drift but will have little effect on anything more substantial.

A light breeze moves at 5–8km/h (3–5mph) and can be felt but will not stir anything but the lightest objects.

Light branches will move in a wind of 8–13km/h (5–8mph), but the tree or bush as a whole will not sway.

Dust is stirred up by a wind of 13–19km/h (8–12mph), which will also blow pieces of paper around.

Small trees will sway in a wind of 19–24km/h (12–15mph).

It is possible to estimate wind speed from heat shimmer using a powerful spotting scope. If there is sufficient heat, ripples can be seen in the air when magnified. Slow ripples indicate a light wind; larger ones imply a more powerful wind. Heat also affects ballistics; warmer air is less dense and creates less drag, which means the round will fly a little further or impact higher up the target. Cold conditions have the opposite effect.

Wind strengths may vary at various heights and distances between the shooter and the target, which further complicates matters. Gusty conditions can make an accurate shot impossible, and may not be due solely to wind conditions. Wind is funnelled down narrow spaces and eddies around the corners of buildings. It may be that an apparently straight and simple shot is too unpredictable to attempt, even though the sniper can see

Types of Mirages

Mirages, or 'heat haze' can give an indication of wind speed. If the air seems to be 'boiling', this indicates calm air, or a wind blowing directly towards or away from the sniper.

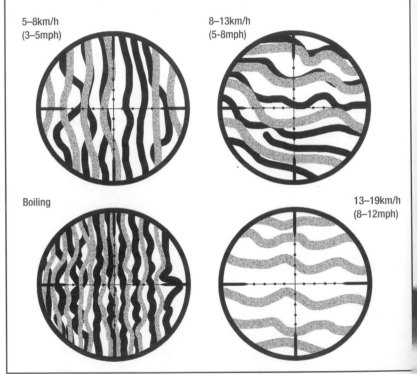

5–8km/h
(3–5mph)

8–13km/h
(5-8mph)

Boiling

13–19km/h
(8–12mph)

directly to the target. The culmination of the sniper's observation is readiness for the shot. He knows the range and the environmental conditions, and can factor these into his aim. Data is recorded on a range card and, now armed with the critical information for a successful firing solution, the sniper settles down to wait for his target.

Clock System

The clock system is used to determine how much of the wind speed to factor into correcting the shot. Wind in the 6 or 12 o'clock position has no effect, 3 or 9 o'clock indicates that the full value must be factored in. From all other angles, the sniper figures half the wind speed into his calculations.

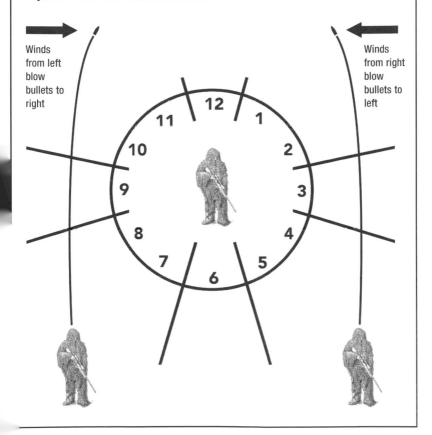

Winds from left blow bullets to right

Winds from right blow bullets to left

Once the sniper team has found a good position, preparing for the shot is a matter of setting up the weapon on a suitable rest, preparing a range card and finally selecting a target. Preparing a range card is a skill that snipers learn in training. It is essential not only in setting up a shot, but also in communicating with an observer or supporting forces such as artillery and air assets. Pre-printed range cards have rings on them representing ranges, centring on the sniper's position. The terrain is sketched in, along with additional data such as air temperature, wind conditions and identifying comments, to allow the sniper and observer to locate an area of interest quickly.

Sketches are also made of terrain features or a more general area. The sniper team may use these during the mission or they can be used for intelligence purposes afterwards. Sketches are not intended to be artistic masterpieces but to convey information and do not need to include fine detail. They show where a given object is relative to others, and its shape and size. It is enough

• •

Hours or even days of moving into position and preparing a hide can be wasted if the sniper does not hit a valuable target. Each shot must be carefully judged to cause the maximum disruption to the enemy.

11

Tactical preparation requires careful observation of the surrounding area and selection of a suitable target.

Tactical Preparation

Data Book

A sniper will record his practise shots on data cards, which are compiled into a data book. He will eventually amass a wealth of information on the performance of his rifle under varying conditions of temperature, humidity, range, wind and a cold or warm barrel. Armed with this information, the sniper can calculate what corrections he needs in order to hit a given target under prevailing conditions.

to represent a hut and add a comment in the margin that it is a radio installation; a perfect depiction of the antennae is not necessary.

A topographical sketch resembles a range card in many ways, but lacks range circles and is used to represent a fairly large area. It does include distances between terrain features as well as notable objects, with any additional comments noted in the margins. A panoramic sketch represents a particular feature and also includes comments on heights and sizes of objects, distances and any other relevant observations, such as the composition of structures or their likely function.

Setting Up the Shot

A sniper can always use his body to support his weapon, but ideally he will be able to rest it in a suitable location. For urban sniping in a fixed location, a tripod rifle mount not unlike a camera tripod can be set up. This is less likely to be possible in the field, however.

Some rifles come with a bipod or monopod attached, and this can be rested on the ground or any suitable support. A bipod or tripod can also be improvised by tying crossed ski poles, sticks or any other straight-ish objects together. Materials for an improvised tripod are usually available in the target area, making it unnecessary to carry one.

The weapon can also be rested on whatever support presents itself, but a skilled sniper will always remember that the best position for shooting is not necessarily the best place to shoot from. That is, an excellent support at just the right height may be available, but it may make the sniper too obvious. It may be counterintuitive to stand back from a

Range Estimation Exercise Scorecard

The only way to become effective in skills such as range estimation is to practice endlessly. Success in an exercise is as important to building the sniper's own confidence as it is a demonstration of his competence to his trainers.

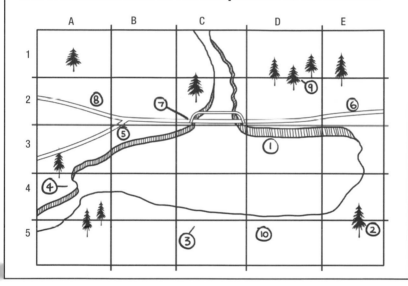

good support and instead shoot from the shoulder over the top of it; however, doing so can make the sniper less likely to be spotted.

There are often good reasons for not resting a weapon on the most obvious support. Most infantrymen are unlikely to note good sniper positions habitually, but once they are fired upon, they will look first in the most obvious places, which generally means the sort of places that a typical infantry soldier would use. Anywhere that offers some cover or concealment and a good field of fire is a candidate, and a sniper does not want to shoot from the obvious place.

Using a Tripod

A tripod is a very stable platform, but requires fairly level ground to use. It is easier to find somewhere to place a bipod or monopod than a tripod.

Cradling the Rifle

In this position, the sniper can make microscopic adjustments in his aim point by tightening or loosening the grip of his left hand on his own arm, which will bunch or loosen the forearm muscles. He can also alter the angle of his left ankle to raise or lower his supporting knee.

Zhang Taofang

Zhang Taofang served in the Chinese Army during the Korean conflict. As the fighting became static in the latter part of the conflict, Taofang was assigned as a sniper to harass UN positions. His issued weapon was only marginally suitable – an old Mosin-Nagant rifle with no telescopic sight. His initial performance was dismal and nearly resulted in his being killed by return fire. However, he learned from his mistakes, picking his shots instead of firing at anything that moved. He also learned how his shot would fall if he aimed at landmarks he picked, and used these to set up lethal shots at UN personnel. In a more dynamic combat environment he might not have been so successful, but by firing from the same spots against static positions, he was extremely effective.

A skilled sniper may pass up the obvious excellent shooting location in favour of somewhere less likely to be immediately targeted. This is extremely important if the enemy has counter-snipers in operation; a good sniper will quickly locate and eliminate lesser snipers who shoot from a predictable position.

Target Selection

There are times when a sniper will aim and shoot at the first target that presents itself, and other occasions when he will decline to shoot at anything except the focus of his mission. This is the case when conducting counter-sniper operations. Other targets may be available but engaging them will compromise the sniper's main mission, which is too important to permit such distractions. An enemy sniper who becomes aware that a rival is in the area will be even more vigilant than before, and may decide either to slip away or to begin hunting the hunters.

A sniper who is intent on eliminating a high-value asset such as a senior enemy officer will similarly pass up inviting but lesser targets. The effort that goes into placing the sniper where he can make his shot is enormous, especially when the target is one that does not often become vulnerable. The sniper will not throw away his chance at the big prize just

Sniper in the Roof

A house with a couple of tiles removed to make a firing position might be obvious, but in an area where buildings have been damaged by recent fighting, there may already be gaps. A few additional ones will not be noticeable.

to take out some random targets of opportunity; not when he has a chance to make a real difference to the course of a conflict.

Order of Importance

Under more routine conditions, targets are selected in order of importance and the level of threat they pose. The level of threat may refer to the sniper team or to friendly forces in the area, and may be present or possible. For example, a dog team is a serous threat to a sniper team as they attempt to break contact, but if it is currently distant and not actively searching, then it is a possible future threat. The observer directing mortar fire on to pinned-down friendly troops is an immediate threat, but not necessarily to the sniper team.

Targets are selected according to the tactical situation. Firing past or over relatively close hostiles may alert them, causing them to become an immediate threat to the sniper team. The probability of a hit must also be considered. A given target may currently be available, but once alerted may have hard cover nearby, which will make a hit virtually impossible. This may move that target up the list a few places or remove it from the list entirely – a target that is not easy to hit once alerted, and who is not critical enough to be targeted with the first shot, may have to be disregarded.

Target Detection Exercise Scorecard

Observation exercises train snipers to spot small objects in a cluttered environment. The same skills can be used to locate a suitable target, to provide detailed reconnaissance or to save the sniper's life if there is an enemy sniper concealed in the area.

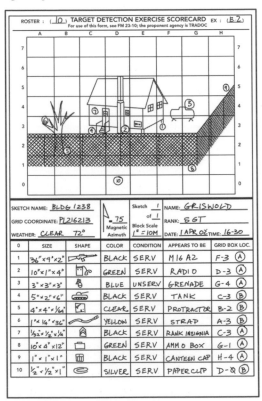

ROSTER : (10) TARGET DETECTION EXERCISE SCORECARD EX : (E 2)
For use of this form, see FM 23-10; the proponent agency is TRADOC

SKETCH NAME: BLDG 1238
GRID COORDINATE: PL216213
WEATHER: CLEAR 72°

Magnetic Azimuth .75

Sketch 1 of 1
Block Scale 1" = 10M

NAME: GRISWOLD
RANK: SGT
DATE: 1 APR 02 TIME: 16-30

0	SIZE	SHAPE	COLOR	CONDITION	APPEARS TO BE	GRID BOX LOC.
1	36"x4"x2"		BLACK	SERV	M 16 A2	F-3 Ⓐ
2	10"x1"x4"		GREEN	SERV	RADIO	D-3 Ⓐ
3	3"x3"x3"		BLUE	UNSERV	GRENADE	G-4 Ⓐ
4	5"x2"x6"		BLACK	SERV	TANK	C-3 Ⓑ
5	4"x4"x1/6"		CLEAR	SERV	PROTRACTOR	B-2 Ⓑ
6	1"x 1/4"x36"		YELLOW	SERV	STRAP	A-3 Ⓑ
7	1/32"x 1/2"x 1/4"		BLACK	SERV	RANK INSIGNIA	C-3 Ⓐ
8	10"x 4"x12"		GREEN	SERV	AMMO BOX	G-1 Ⓐ
9	1"x 1"x1"		BLACK	SERV	CANTEEN CAP	H-4 Ⓐ
10	1/2"x 1/2"x1"		SILVER	SERV	PAPER CLIP	D-Q Ⓑ

Target Priorities

It is unlikely that a sniper, however skilled, would be able to eliminate an entire patrol. He can, however, render the force incapable of coherent action with just one or two shots at the right target.

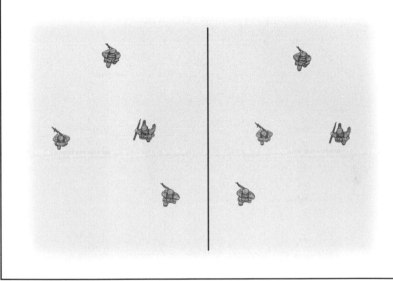

The sniper must also consider the effects of eliminating a given target on the enemy and on friendlies. Some targets are sufficiently isolated from support that it may be possible to eliminate them without alerting their comrades, especially if suppressed weapons are in use. With all this in mind, the sniper needs also to consider how the enemy will react to a shot, and the effects on his

mission. As already noted, a mission to eliminate a high-value asset should not be compromised by attacks on lesser targets.

An observation mission may be of greater value than shooting an enemy gunman or two, so the sniper may have to make the difficult decision to do nothing with a perfect target in sight. Likewise, a shot that might result in an enemy reaction against

Chuck Mawhinney (1949–present)

Chuck Mawhinney is credited with 103 confirmed kills during his career as a US Marine Corps sniper. According to the recording convention at the time, to be confirmed, a kill had to be witnessed or personally checked by an officer, which was not always possible. Thus Mawhinney also had more than 200 'probable' kills to his name that were not confirmed. These included 16 members of an enemy patrol killed with his M14 combat rifle rather than his sniping weapon. Each one was killed by a headshot, but the feat was not witnessed and thus is not part of the official record.

nearby friendlies needs to be carefully considered. The sniper needs to remain aware of the big picture and make a shoot/no-shoot decision accordingly.

Valuable Targets

Where possible, a sniper's fire will be directed against the most valuable members of an enemy force, or against those that pose the most imminent and severe danger to friendlies. An otherwise ordinary insurgent who has obtained a flanking position, where friendly infantry cannot hit him, may become a key target, if eliminating him improves the tactical situation for the force as a whole or prevents casualties. The following are valuable targets:

Snipers are the absolute top priority under most conditions. The chance to shoot an enemy sniper comes rarely, and the damage he can do is enormous. Therefore, unless there is a pressing reason not to, a sniper will eliminate any enemy counterpart he can identify. This is also a matter of self-preservation, since the best weapon against a sniper is another sniper.

Senior commanders capable of co-ordinating effective enemy forces are a high-priority target due to the disruption the enemy will suffer. Their immediate staff are also often key targets, as they are likely to oversee such critical functions as intelligence and logistics. Eliminating any of these personnel will make the enemy less effective for a time.

Dogs and handlers are more likely to detect snipers and to maintain pursuit as they exfiltrate than any other enemy forces. The handler is a higher-priority target than the dog, under most conditions, as an undirected dog may confuse the enemy more than help them.

Scouts and artillery observers are, like snipers, a serious threat to a sniper team and to friendly forces. They are skilled and usually experienced personnel whose capabilities go beyond those of the typical enemy combatant. The loss of effective observers and scouts can cripple an enemy force in the longer term and will reduce its ability to respond to the sniper in a more immediate time-frame.

Command personnel such as officers and non-commissioned

Target Selection

A sniper can strike at the enemy even in supposedly safe areas. He must choose where to place his shots for best effect – personnel will take cover, but inanimate objects will remain where they are unless moved. A sniper might select the tyres of an enemy vehicle, to hinder enemy mobility, or aim to puncture fuel tanks.

M24 Sniper System

The US M24 sniper system was scheduled for replacement by the M110, a semi-automatic weapon. However, many examples have been upgraded which suggests that the US Army intends to retain the M24 for some time to come.

officers, including vehicle commanders and ordinary infantrymen who start to display initiative and leadership, are key targets in the immediate sense. Eliminating them will weaken the enemy response and can make it much easier for conventional friendly forces to inflict a defeat. In the longer term, the loss of experienced officers and NCOs lessens the quality of the enemy force.

Specialists such as communications personnel, support weapon crews, mechanics and drivers require more training than typical infantrymen. In the immediate sense, their loss can weaken an enemy force's ability to call in support, to manoeuvre or to fight. The loss can be obvious, such as silencing a machine gun, or a more subtle loss of capability, such as the lack of accurate mortar support or rapid evacuation from the position.

Equipment such as vehicle optics, communications and radar sets, and many weapon systems, can be crippled by a bullet strike. The loss of such equipment can be immediately felt, for example in the loss of air-defence radar or, longer term, a lack of communication leading to an operation becoming dislocated. Either way, the sniper can influence wider events with a single shot.

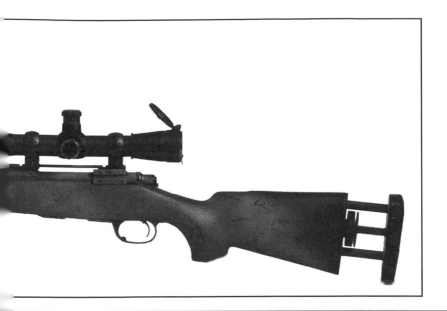

Timothy L. Kellner

Timothy Kellner is credited with 139 kills during his service with the US Army in Iraq. Much of his work was in urban terrain, where it was not always possible to confirm a kill or even find the target's body, so Kellner has additional 'probables' to his name – possibly as many as 300. Most of Kellner's sniping was done with an M24 Sniper System, which is based on the bolt-action Remington 700. Like many modern snipers, Kellner operated in an urban environment where targets were hard to identify and there were always noncombatants to consider. Although the requisite skills are the same, this environment is significantly different to that found in an open-country war zone.

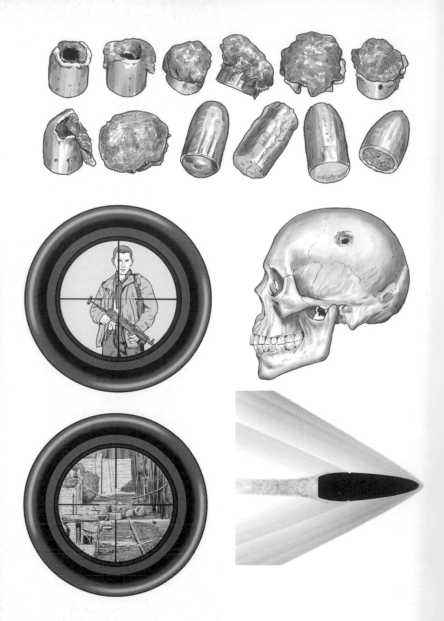

The majority of snipers' shots are directed against human targets and will kill or cause grave injury. However, this is not the main purpose of shooting. In some ways, a sniper's shot is a very personal act – he selects a target and then shoots to kill. But the kill is usually a means to an end.

Even when the mission's aim is to eliminate a given individual such as a general or an enemy sniper, it is not his death that is sought but the effects that it will have on the enemy.

The sniper kills another sniper to prevent him shooting friendly personnel, or eliminates an enemy officer to stop him playing an effective part in operations. He need not – and usually does not – hate or even dislike his target. He will, however, have a more graphic view of the effects of his shot on the target than most soldiers.

He may be looking right at the target when he is killed or critically wounded, and must be psychologically prepared for this. If he is unable to deal with his feelings on killing another human being in cold blood, then his career will be short and probably tragic.

. . . .

To achieve the greatest results a sniper must understand exactly what he is about to do to his target, and be willing to shoot regardless.

12

A single shot can have a profound effect, not just on the victim but on the military machine that he is a part of.

The Effect of the Shot

Dragunov SVDS

The Dragunov SVDS is a compact version of the rifle, intended for use by paratroops and other personnel who may have to move through confined spaces.

Gunshot Wound

Bullets typically make a small entry wound and a much larger exit wound. A 'through-and-through' will generally cause lesser injuries than a round that stops in the body and imparts all of its kinetic energy to the target.

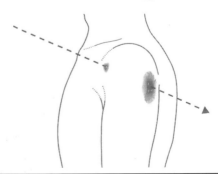

It uses a folding stock and more advanced materials than the original, but it is essentially the same weapon.

Bullet Wounds

Most people do not really understand how bullets cause injury. Movies and television are driven by the needs of a plotline or dramatic effect rather than physics; bullets have the effects that the producers want them to have. This creates a false impression of the capabilities (or lack thereof) of many weapons.

In extremely simple terms, there are two different but related effects that firearms can cause: wounding and stopping. 'Stopping power' is the ability of the weapon to make the target stop immediately, whatever he is doing. It may or may not be connected to lethality or wounding capability – for example, less-lethal 'beanbag' round may put an assailant on his back, stunned and most definitely stopped, but causes relatively little wounding and is unlikely to be lethal. Similarly, a small-calibre bullet from a handgun might not immediately stop an opponent, but may cause him to bleed to death over an extended period.

For some combatants, such as the lone police officer facing a group of hostiles or a soldier caught up in a close-range firefight, stopping power is vitally important, as it increases his chances of survival. If an opponent inflicts a return wound, then expires, this is not much of a win. On the other hand, snipers (and assassins) are more concerned with lethality than stopping power much of the time. Whether the target dies immediately or later may not matter all that much; a critically wounded but still-functional target may not be a threat if he is several hundred metres away.

Shoot to Kill

However, snipers normally shoot to kill. As a rule, most casualties of gunshot wounds who reach suitable trauma aid will survive, and a downed but conscious hostile may still be able to accomplish his task. A sniper might choose to wound a target for various reasons, perhaps to draw others out of cover, but it is most common to shoot to kill.

There is a correlation between

wounding capability and stopping power for most weapons – a more potent weapon will normally be both more lethal and have better stopping power – but it is also possible to use wounding capability to ensure a 'one-shot stop'. A severe wound to a non-critical area such as some parts of

Shooting a Moving Target

A target that is moving directly towards or away from the shooter requires little lead. Targets that are moving laterally must be led more, depending on their speed and the angle of movement.

the torso or leg may not succeed in stopping the target, but the same level of injury directly to the heart or brain will kill him instantly. This last factor is particularly important when trying to eliminate key enemy personnel or a criminal who is holding hostages.

Body Shot

Any penetrating injury on the torso is potentially life-threatening, and the centre of body mass is a much easier target than the head. However, a body shot does not always guarantee an immediate kill.

Psychological and Physical Effects

It has been noted that the effects of a bullet wound are partially psychological. To put that more simply, 'any hit that hasn't already put you down, isn't going to put you down'. Someone who is going about their business and is not in a 'fight mentality' frame of mind will probably fold up and go down if shot, even if the injury is not very severe. But someone who is fighting or running for their life when they receive a non-critical gunshot wound will probably carry on, if they physically can. They may collapse afterwards, and may even die, but the same level of injury will produce different reactions depending on the mindset of the target.

However, physical effects trump psychological ones. The destruction of a vital organ or the breaking of a bone will always have the same effect, regardless of the target's toughness or mindset. Therefore, a sniper shoots for a physical effect, which is predictable and certain, rather than hoping to inflict a wound serious enough to incapacitate the

Types of Cavity

A bullet causes different types of cavity as it passes through the body. The 'crush cavity' is caused by the physical passage of the bullet and is the same diameter as the bullet, which may expand in the wound. The 'wound cavity' is created as tissue surrounding the bullet's passage is destroyed.

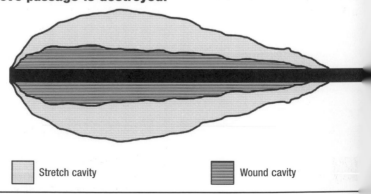

Stretch cavity

Wound cavity

target. Physical effects are, of course, greater with more powerful weapons, i.e. those that shoot a heavier bullet at a greater velocity.

Science of Ballistics

A bullet causes injury as a result of its mass and energy, but this is not a simple matter. How efficiently the bullet transfers its energy into the target, and what parts of the body are affected, are important factors. These in turn are affected by how the bullet performs when it strikes tissue and bone. A sniper needs a working

knowledge of these factors if he is to achieve the right results or predict the effects of a given shot.

The science of ballistics deals with projectiles such as bullets, and within the overall field there are three subfields: internal, external and terminal. Internal ballistics deals with the bullet inside the weapon and external ballistics is concerned with its motion after it is fired. The latter is what is normally referred to when the term 'ballistics' is loosely applied. Terminal ballistics is the study of a bullet's behaviour once it strikes something denser than air and includes the even more specialized area of wound ballistics.

Wound ballistics indicates that a bullet causes injury in three different ways:

Laceration or crushing is caused when the bullet directly strikes tissue, organs or bone and carves a path through them. Relatively low-velocity projectiles such as bullets from most handguns cause injury primarily through these mechanisms. This means that, as a rule, a handgun round that does not strike a vital organ or sever a critical blood vessel may cause relatively little damage. This makes handguns more survivable than higher-velocity weapons.

Shockwaves are created by high-velocity projectiles such as those from rifles, and compress the soft

The 'stretch cavity' is temporary, and is caused by the energy imparted to tissue by the bullet. This cavity collapses quickly and may not be evident after the event.

Crush cavity

Bullet Impact

The skull is very hard but can be fractured by the forces it is subjected to in a bullet impact. Thus there may be damage to the skull in other areas than the bullet impact point.

Lacerated skulls

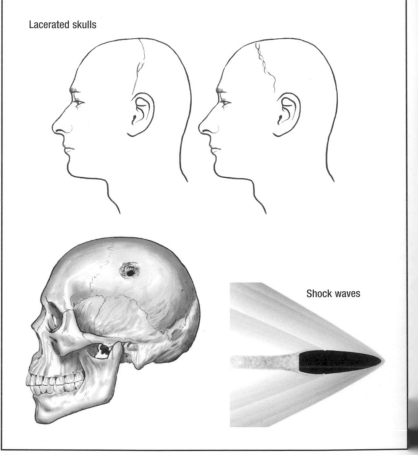

Shock waves

tissues ahead and to the sides of the bullet. Human tissue is mostly water, i.e., it is a dense fluid medium, which transmits shock.

This has led to various myths about hydrostatic shock, notably the notion that there are weapons so powerful that being shot in the toe will cause the target's heart to explode. While this is not the case, shockwaves from high-velocity bullets passing through soft tissue do cause extremely significant levels of injury.

Cavitation is caused when a bullet imparts some of its kinetic energy to the target's tissues, setting them moving away from the bullet. The resulting wound cavity is short-lived – just microseconds in most cases – but the resulting damage can be very severe.

Tissues can be severely damaged as the cavity opens and then collapses again. This phenomenon is sometimes confused with the wound track or permanent cavity left behind when a bullet pushes through tissue and leaves a hole. In this case, the tissue is permanently displaced, often by being blown out of the body through the exit wound.

Bone fragments A bullet that strikes one can be deflected or may fragment, creating multiple wound tracks. Bone fragments can also act as secondary projectiles that will cause further damage. Bones can be broken even if not directly struck, largely due to the cavitation phenomenon. The skull can in some cases deflect a bullet, but this normally only occurs with lighter, lower-velocity rounds.

A bullet that enters the cranium, but has insufficient energy to punch out the other side, may bounce around inside the skull, which causes even more brain trauma than a through-and-through shot. Skull fragments and perhaps pieces of debris from a helmet can also injure the brain.

A Sure Kill

The most serious injuries inflicted by gunshot wounds are those to vital organs or major blood vessels such as arteries. Bone and tissue injury may be survivable, and relatively minor bleeding from a wound to a non-vital area can be reduced to the point where the casualty will survive for a lengthy period; if this is long enough to reach suitable medical treatment, then a full recovery is possible and survival highly likely.

To be sure of a kill, therefore, a sniper needs either to deliver a significant wound to a vital organ, or cause sufficient trauma from secondary injuries – cavitation, damaged blood vessels and bone fragments – to lead quickly to a fatality. The characteristics of the

Kill Shots

A shot to the brain will normally be instantly fatal. It was long believed that hitting the medulla oblongata, a small area at the base of the brain, was necessary to prevent a gunman from reflexively firing his weapon, but there is evidence that any brain hit will suffice.

Head

A hit to the heart will also cause death, either instantly or from rapid internal bleeding.

Chest

Bullet Flight

A bullet that *yaws* in flight will lose velocity quickly and may be inaccurate, but may possess superior wounding characteristics as it will almost certainly yaw after it has hit the target. This causes a more rapid dump of kinetic energy into the target and a wider wound track.

A spinning bullet will *precess* about its long axis as a result of the complex forces acting upon it. The nose of the round points generally in the direction of flight, while its tip describes a circle. A precessing bullet is inherently more stable than one that is yawing in flight.

Nutation, or 'nodding' is a complex movement which can cause a bullet to tumble if it becomes excessive. It is more of a problem with long, heavy bullets than shorter and lighter rounds, but tends to self-correct due to the gyroscopic stabilization imparted by spinning the bullet.

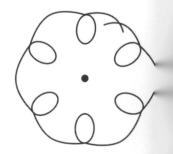

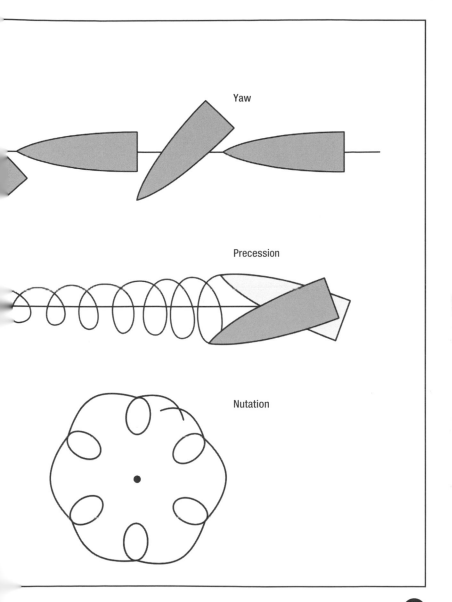

Yaw

Precession

Nutation

Bullet Impact

Most bullets will expand or 'mushroom' to some extent as they decelerate rapidly after entering the target. Some fragment, which creates several smaller wound tracks, often at widely divergent angles. Softer and more blunt-ended bullets expand most readily. Ammunition specifically designed to expand in the wound is illegal for use in international warfare but is permitted for hunting and law-enforcement use.

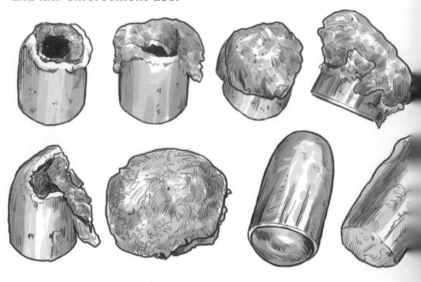

bullet, as well as its placement, will determine the fate of the target.

Bullet Dynamics

There is a misconception in some quarters that some rifles shoot bullets that tumble in the air, making them more lethal. This is not the case; a 'tumbling bullet' of this sort would be extremely inaccurate and shed its energy fast, making its range rather short. Bullets are spun to make them

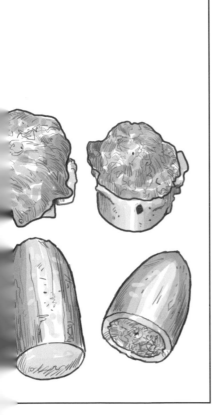

Bullet Shape and Composition

A bullet must be aerodynamic and also stable in flight when spun, if it is to be accurate. It must also be dense and heavy. Lead is an ideal material for bullet manufacture, not least since it will deform on impact, which can increase the level of injury inflicted on the target. Some bullets are sheathed in copper or other materials, or have a core of a different material to alter their flight and armour-penetration characteristics.

stable in flight, whereas a tumbling round would be anything but stable.

A bullet typically has a pointed nose and a long axis, about which it is spun. In flight it can move in three axes: pitch, roll and yaw. All bullets roll (spin) about their long axis as a result of rifling, but they also pitch and yaw in flight. Pitch refers to the up-and-down movement of the nose of the bullet relative to the long axis; yaw is side-to-side movement.

Bullets do not fly straight, nose-first, but move slowly around their long axis in flight (precession). This means that the bullet is not always pointed exactly forward as it flies; the nose will be slightly off the axis of the bullet, moving around in a circle centred on the direction of flight. Thus, the bullet may not hit the target exactly nose-first, even if the target is square on to the direction of flight – and that is by no means guaranteed. When a bullet hits flesh it slows down rapidly, which may cause it to yaw or

Bullet Shape

The typical bullet is of a round-nose type, but there are specialist versions for various applications. Wadcutters are primarily used for target shooting, though they do offer good stopping power for self-defence.

Round nose

Semi-wadcutter

pitch end-for-end in the wound. This is the origin of the 'tumbling bullet' myth. The result is that the bullet will end up driving though flesh backwards, which increases the amount of wounding inflicted. This will not occur in a shallow, grazing wound, but a round that punches into the torso may have time to tumble before coming to rest or exiting.

A bullet moving blunt-end-first will shed its kinetic energy faster than a pointed round that can pierce flesh more easily. This, in turn, leads to

The semi-wadcutter is a compromise between the better ballistics of the round-nosed bullet and the stopping power of a wadcutter. Jacketed hollow-point rounds are designed to expand upon impact, but have a hard jacket to improve ballistics and to prevent soft lead being deposited in the barrel during firing.

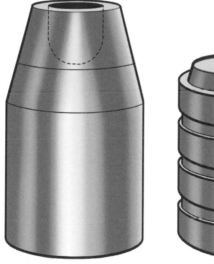

Jacketed hollow-point

Wadcutter

Bullet Fragmentation Wound

A round that fragments in the wound will impart a great deal of force to nearby tissue, displacing some of it, rather than carrying its energy through and out of the body. Small bullet fragments can tear blood vessels or damage organs by direct impact, though they cause little cavitation.

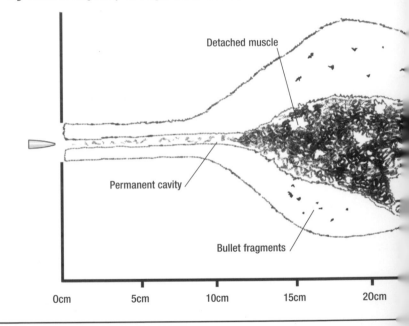

Detached muscle

Permanent cavity

Bullet fragments

0cm 5cm 10cm 15cm 20cm

greater wounding effects. The round may also fragment; smaller parts slow down more quickly and impart their energy to nearby tissue, causing larger wounds. Fragments also increase the chances of striking a vital organ or blood vessel.

Enhanced Lethality

Highly stable rounds are the least likely to yaw in the wound, meaning that the most accurate ammunition may cause lesser injuries than ordinary bullets. There are various ways of increasing bullet lethality

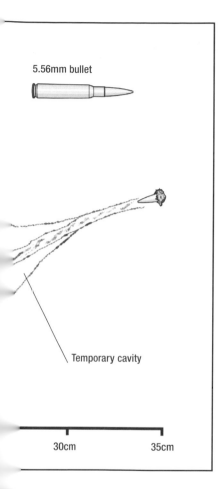

5.56mm bullet

Temporary cavity

30cm   35cm

Expanding Bullets

A bullet made of very soft lead will expand or 'mushroom' on impact, which makes it shed its kinetic energy fast and cause greater injury. Expanding bullets also widen the wound track, which enhances the degree of tissue trauma suffered. Another way to cause a bullet to expand on impact is to give it a hollow tip or an X-shaped groove at the tip, both of which cause expansion. Hollow-point rounds are somewhat prone to fragment, which may be useful in some cases but does reduce penetration of light cover or body armour.

The full metal jacket used in military ammunition makes bullet expansion less likely, although jacketed hollow-point rounds will expand in the wound. Semi-jacketed ammunition leaves the soft tip of the round exposed, balancing penetration with expansion. Expanding rounds of this type are sometimes referred to as Dum-dums after an arsenal in India used by the British Army in the Nineteenth Century, while experimentation with expanding bullets was under way. Expanding bullets were banned for use in warfare by the Hague Convention of 1899, but remain legal for self-defence and law-enforcement, hunting and the like.

Armour-Piercing

Armour-piercing ammunition is

and, of course, a precise enough shot – if aimed at the heart or head will always be lethal. Some enhanced-lethality rounds are controversial: some are banned in international law, while others are entirely legal.

BMG Round

The 12.7mm (0.5in) BMG (Browning Machine Gun) round is used in some anti-materiel rifles as well as heavy machine guns. It is large enough to be useful as the basis of specialist ammunition such as explosive or incendiary rounds.

designed to penetrate hard surfaces, which by definition makes it more effective at moving though flesh as well. While an armour-piercing round that strikes an organ is as likely to be fatal as any other bullet, it sheds its kinetic energy less readily and is thus less likely to cause serious wounding if it does not strike an organ. The trade-off is worth it if the sniper must engage a hard target such as an armoured hostile, or shoot through an intervening object. Armour-piercing rounds are also used for anti-materiel work.

There are various ways to enhance penetration. Bullet shape is a factor – a more pointed round will penetrate better, so long as it strikes in a reasonably perpendicular manner. A bullet made of harder material will

retain this shape better upon impact, which enhances penetration. Many armour-piercing rounds use a penetrator core formed from a hard, dense material such as steel or tungsten carbide. The penetrator is encased in a copper or cupronickel jacket to protect the barrel of the weapon from excessive wear.

Very large-calibre weapons can use discarding sabot rounds, which work much like anti-tank ammunition: a needle-like penetrator is encased within a sabot ('shoe'), which is jettisoned in flight. Other options include armour-piercing-incendiary and high-explosive-armour-piercing ammunition. Although not intended for anti-personnel use, these rounds would do immense damage to anyone they hit, and might be used

Estimating Hold-off

The mil dots on a sniper's scope can be used to measure the correct hold-off distance to compensate for wind or bullet drop on a long shot. Estimating how much hold-off is required, and in what exact direction, is an art, but measuring it out with the scope is a precise science.

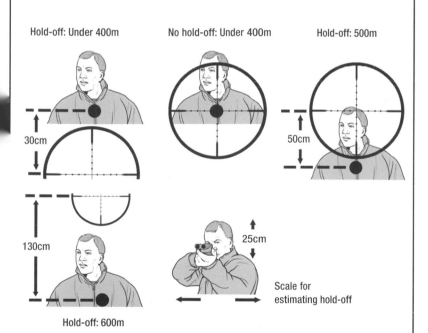

Hold-off: Under 400m

No hold-off: Under 400m

Hold-off: 500m

30cm

50cm

130cm

Hold-off: 600m

25cm

Scale for estimating hold-off

Five O'Clock Charlie

During the Vietnam War, US troops positioned near Da Nang came under fire at the same time every day by a particularly persistent enemy sniper who became known as 'Five O'Clock Charlie'. Various attempts to eliminate the sniper having failed, the troops were in the process of moving a 106mm (4.17in) recoilless rifle up to the top of a hill when the realization dawned that this sniper had not actually managed to hit anyone. Rather than blast his suspected position with 106mm high-explosives, it was decided to leave him alone in case he was replaced by someone competent. For a time the troops amused themselves by hoisting 'Maggie's Drawers' – a red flag used to signify a miss on the firing range – every time he fired. Five O'Clock Charlie fired several shots at the US positions every day for two months, and never hit anything.

needed. Extremely long-range rounds are also available, which are optimized for sniping at distant personnel targets, where enhanced penetration is not required.

Mortal Danger

Being shot at is, in the words of one British officer, 'a thoroughly disagreeable experience'. Nobody actually likes being in mortal danger, but there are some threats that unsettle soldiers far more than others. Coming under artillery fire without being able to reply is galling, but mortaring is often considered to be worse. Artillerymen are a long way away, firing shells up in the air towards a map grid. Mortar crews, on the other hand, will be quite nearby, which makes their attacks somehow more personal.

Similarly, the rifle and machine-gun fire exchanged during infantry combat is more tolerable than sniping. Much of the shooting that goes on is aimed in the vague direction of a muzzle flash or a moving hostile.

Someone who makes a good target may be singled out, but if he takes cover for a while, the enemy will more than likely start shooting at someone else. This sort of combat is impersonal – it is one group against another. Besides, th vast majority of shots miss their target. Basic infantry skills such as making use of cover, movement and suppressive fire can greatly enhance the safety of any given soldier, and his confidence in those skills will enable him to take calculated risks in order to get the job done.

The Growing Fear

Snipers are different. A sniper will hunt an individual and rarely misses. He may strike at any time, from a concealed position some way off. The ordinary soldier often has no means to respond to this threat, and that is a most demoralizing position to be in. Even in fairly low-threat areas there is always the suspicion that a sniper may, even now, be drawing a bead. This sensation becomes much worse once the realization dawns that a given shot came from a sniper. The suspected threat becomes imminent, and fear grows.

A lone sniper can hold up a fairly large body of troops for some time. There have been cases of companies or even battalions halted for lengthy periods by a few shots. This has much to do with the perception of risk. If a unit comes under fire and most of that fire is ineffective, then the risk perceived by any given member of a unit is relatively low. The troops will usually be willing or possibly eager to mount an attack,

even if this means moving out of cover to begin a fire-and-movement advance.

The precise and personal nature of sniping alters this perception. Each soldier knows that if he draws the sniper's fire, he will probably be hit. This 'knowledge' may be incorrect; snipers are by no means infallible, and indeed there may be no sniper involved. It is possible that a sniper's presence could be inferred from a couple of lucky hits. What matters, though, is what the troops under fire believe, not the actual circumstances.

If the members of a unit come to believe that they are very likely to be picked off if they break cover, then they may be entirely unwilling to do so, especially if they cannot see where the shots are coming from. Those that might have led a counter-attack may be killed or may choose not to say 'follow me' if this seems likely to lead to being shot.

Sufficiently desperate circumstances may counterbalance the perceived risk, but as a rule the involvement of a sniper raises the perception of personal risk and may dissuade the unit from doing more than taking cover and calling for support.

There are times when a sniper can make use of this phenomenon to pin down an enemy unit or effectively deny an area to the enemy. If a sniper has killed several friendlies in a given

Street Target

Under the magnification of a sniper's scope, small details become easy to make out and a supposedly concealed hostile can often be quickly spotted. What the sniper can see, he can shoot.

area, others will become reluctant to enter it. They may still do so, but their caution may inhibit operations. This effect lasts after the sniper has moved on, at least for a time.

Indeed, a sniper could move between several areas of a city or battle zone, influencing one by his actual presence and the others simply by the suspicion that he may still be there.

Conversely, sometimes it is worth disguising the presence of a sniper. A single shot followed by a casualty, with no enemy in sight, paints a pretty obvious picture. The troops who were attacked will react with anti-sniper drills and may call in assistance such as a sniper of their own. However, if they are engaged with an obvious enemy force, they may not realize that some of their casualties are being inflicted by a different source.

A conventional infantry position or security checkpoint can even be used to draw out an enemy force into a sniper ambush. If the enemy knew that there were multiple snipers covering the approaches to their target, they might decide not to attack, but if they thought that all they were facing is a lightly held security position, they might be willing to come out and fight.

Instinct vs Training

Human instinct is to group together for mutual protection. In the days of hand-to-hand combat with swords and spears, this was an effective tactic. Since the development of firearms, 'bunching' has become highly dangerous, as it presents the enemy with an attractive and relatively easy target. Troops are constantly warned not to bunch, but against this must be set millennia of instinctive behaviour.

The high level of threat posed by a sniper will incline many troops to seek the illusion of protection afforded by their comrades and will force them into taking the best cover possible. It will also limit their ability to move or even observe the situation. Thus, a sniper can push together an enemy force and hold it in place, enabling other forces to exploit the situation. This may be air or artillery support directed by the sniper, or other infantry forces. Snipers can also serve as a rearguard for a withdrawing force, slowing down an enemy advance by the threat they pose, while friendlies break contact and take up a better position.

Thus the effects of a single sniper's shot extend far beyond the life taken. A single shot can reduce overall enemy capabilities, weaken morale and even deny a large area to the enemy for a time. All of the sniper's skills are aimed at this one goal – taking a shot that makes a real difference to the overall situation.

FINAL NOTES

The term 'sniper' is misused a great deal in popular culture, along with several related terms. It sometimes seems that anyone with a rifle is a 'sniper' to the media, and all rifles are 'high-powered rifles' or 'high-velocity rifles', even if they are actually low-powered 'plinking' guns. For many people, the term 'sniper' has, to some extent, become associated with criminals and psychopaths; i.e. a 'sniper' is someone who climbs a bell tower and starts shooting indiscriminately.

This is almost entirely the opposite of what a sniper actually is. A sniper is a trained professional who is very selective about who and what he shoots at. After the careful selection and training process he has gone through, a sniper is one of the least likely people to 'snap' and start murdering innocents.

This is just as well, as even an out-of-practice ex-sniper has some dangerous skills at his disposal, should he choose to recall and use them.

What is a Sniper?

So what exactly do we mean by 'sniper'? Obviously, a successful graduate of a sniper-training course is given the title, but the question is a little more complex than that. It seems reasonable to refer to a hostile combatant who is taking careful shots from a concealed position as a sniper, even if he is a self-taught insurgent who uses a standard assault rifle. What he is doing is sniping, even if he is not a qualified sniper. With enough experience he will become a true sniper, although he may never undergo formal training.

Skillset, mindset, sniper

Thus the status of 'sniper' can be considered a combination of a skillset and a state of mind. A sniper is someone who has the right skills and uses them effectively, whether formally trained or not. Indeed, a training school with very low standards might confer the title of 'Sniper' on its graduates, but they would not be snipers in the truest sense – a title is one thing, but in the field they would likely not come up to scratch.

Perhaps the only useful definition of a sniper is someone who can carry out the missions required of a sniper. This includes the whole set of skills, marksmanship and, of course, the appropriate mindset. Those who cannot do the job, whatever title they hold, are not snipers. Those who can however they managed to acquire their skills, are the real thing.

All in the Game?

The mystique of the sniper has inspired several video games, some specifically about snipers and others where sniping is an integral part of the game experience. Some games portray sniping quite realistically; others rather less so. The more realistic games do reflect some aspects of sniping quite well, but they tend to leave out the endless waiting, being cold, wet and half-eaten by bugs. A game based around these facets of sniping might not be all that much fun...

Similarly, some players use real sniper tactics and techniques while others run around shooting with a scoped rifle – there is a world of difference. It is possible to make

•••••••••••••••••••••••••••••

A US special forces soldier aims a Mk 12 sniper rifle during Operation 'Iraqi Freedom', September 2007.

The bolt-action M40A1 is one of the most reliable sniper rifles in service today.

US Marines from the 24th Marine Expeditionary Unit practice camouflaging techniques in Djibouti, Africa.

....

effective use of sniping technique in a game, but ultimately games are intended to provide entertainment and diversion. Real-world sniping is rather different.

This author's attempt to try out sniper technique in a multiplayer game resulted in an experience that illustrates the problems of working closely with less subtle comrades and also perhaps the difference between a video game and a real conflict. The incident inspired the following poem:

He emptied a magazine down at the foe,
As the firefight raged in the alleys below.
He was braver than most men would normally be,
The team-mate who came and stood next to me.

There I was sniping, face down to the ground,
Shifting position after firing each round.
'No need to take cover!'
– I didn't agree.
But so thought the team-mate who stood next to me.

With all of that shooting, the full-auto kind,

*For enemies searching, we weren't
hard to find.
Still he chalked up some hostiles –
he got two or three,
The ham-fisted hero that stood
next to me.*

*Then an enemy soldier made his
daring raid,
Running towards us with ready
grenade.
One a five-second fuse that he'd
cooked down to three.
It flew past my team-mate and
landed on me.*

*My team-mate, whose fault it was,
fled down the block,
But I was too close and I caught
the full shock.
So I don't ask for much, but I will
make this plea:
If you're going to draw fire, then
don't stand next to me!*

Autonomy, Uniqueness, Intelligence

Facetious as this poem may be, there is a point to be made here. The absolute last thing a sniper needs is to attract the attention of the enemy, or to have it attracted in his direction. If he is to do his job properly, then he needs to be given sufficient autonomy to make his own decisions, and allowed to make best use of his skills away from other troops. This, in turn, requires that he be trusted rather than micromanaged by his superiors.

A sniper is a potent military asset, but one that cannot be closely controlled without wasting much of his potential. He needs suitable equipment and training, but he must also be placed under the command of someone who understands his unique capabilities and his limitations. He can do things that ordinary soldiers simply cannot, but he cannot deliver a miracle on demand every time.

The days of massed infantry armies that had to be kept under close control are gone. Technology provides many force-multipliers that enhance the effectiveness of each soldier, which is increasingly necessary in a world where armies are stretched thin in manpower-intensive, low-intensity conflicts. Being able to field 'thinking soldiers' is also a massive force-multiplier. Anyone can obey an order, but an intelligent soldier can decide for himself how best to support his comrades and improve the tactical situation without needing to be told.

Perhaps the reason why sniping was neglected after each major war of the Twentieth Century was that snipers do not fit neatly into the perfectly aligned military machine. Yet, time and again, it has been proven that they are necessary and most effective when given a task and allowed to get on with completing it. Snipers are, perhaps,

the ultimate thinking soldiers, and in an age where military budgets are increasingly squeezed they represent an incredibly valuable investment, but only for those who know how to make best use of them.

The Shot Not Taken
In the final analysis, a sniper is

Legionnaires from *2e Régiment Étranger d'Infanterie* pose somewhere in Afghanistan. The soldier at the front is armed with a PGM Hécate II anti-materiel rifle, his comrade with an FR F2 sniper rifle.

● ●

own, often lonely, world. He sees the effects of his shot more graphically than most other combatants, but must remain willing to go out and do it again.

The sniper kills clinically, hampered rather than enabled by adrenaline, and knows that he is unlikely to be hailed as a hero. Yet he does what needs to be done, quietly, in the shadows, and moves on. Few people have the makings of a sniper, and fewer still qualify. But those who do, wield power far beyond their numbers.

Ironically perhaps, the best way to understand the influence of the sniper is to imagine what might have happened had the sniper not made his shot. The battles never fought, the casualties never taken, the soldiers coming home intact, unharmed by an explosive never planted – these are the results of the sniper's endless patience, his rigorous training and his clinical, expert killing shot. A sniper is someone who is willing and able to do a terrible thing in order to prevent others from doing worse.

someone who has the skills and the mental attributes to make a major difference in a conflict. He is both the hunter and the protector, operating with great patience in his

APPENDICES

TOP SNIPERS FROM THE AMERICAN REVOLUTIONARY WAR TO THE PRESENT

Country	Sniper	Conflict	Confirmed Kill
US	Timothy Murphy	American Revolutionary War	not known
UK	Patrick Ferguson	American Revolutionary War	not known
UK	Thomas Plunkett	Peninsular War	not known
US	Sgt Ben Powell	American Civil War	not known
Canada	Francis Pegahmagabow	American Civil War/ World War I	378
Australia	Billy Sing	World War I	150
US	Henry Norwest	World War I	115
Finland	Simo Häyhä	Winter War (Finland vs Soviet Union)	705
Soviet Union	Lt Vassili Zaitsev	World War II	225
Germany	Pte Matthias Hetzenauer	World War II	345
Germany	Josef 'Sepp' Allerberger	World War II	257
Soviet Union	Lyudmila Pavlichenko	World War II	309
New Zealand	Alfred Hulme VC	World War II	33
China	Zhang Taofang	Korean War	214
US	Adelbert F. Waldron	Vietnam War	109
US	Carlos Hathcock	Vietnam War	93
US	Chuck Mawhinney	Vietnam War	103
US	Frank Grieci	Operation Desert Storm	15
US	Scott Dennison	Operation Desert Storm	14
US	Timothy L. Kellner	Operation Iraqi Freedom	139

RECORD SHOTS

Country	Sniper	Conflict	Rifle	Distance
UK	Craig Harrison	Afghanistan	Accuracy International L115A3	2475m/ 2707yds
Canada	Rob Furlong	Afghanistan	McMillan TAC-50 Tactical Rifle	2430m/ 2657yds
Canada	Arron Perry	Afghanistan	McMillan TAC-50 Tactical Rifle	2310m/ 2526yds
UK	Christopher Reynolds	Afghanistan	Accuracy International L115A3	1853m/ 2026yds
US	Brandon McGuire	Iraq	M82 Barrett .50	1310m/ 1433yds

BEST SNIPER RIFLES FROM THE AMERICAN REVOLUTIONARY WAR TO THE PRESENT

Rifle	Country	Calibre	Effective Range
Kentucky Long Rifle	American Colonies	15.2mm/0.60in	100–250yds
Whitworth Rifle	United Kingdom	11.5mm/0.451in	800–1000yds/ 730–910m
Ferguson Rifle	United Kingdom	17.27mm/0.68in	Variable
Baker Rifle	United Kingdom	11.43mm/0.45in	Variable 100– 300yds/91–270
Mauser Gewehr 98	Germany	7.92mm/0.312in	800m/875yds
Gewehr 41	Germany	7.92mm/0.312in	400m/437yds
Lee-Enfield No.4 Mk1	United Kingdom	7.7mm/0.303in	550m/600yds
M1 Garand	United States	7.62mm/0.3in	440yds/402m
Mosin-Nagant 1891/30	Soviet Union	7.62mm/0.3in	750m/820yds
Karabiner 98K	Germany	7.92mm/0.312in	800m/875yds
Type 99	Japan	6.5mm/0.256in	—
Springfield 1903A4	United States	7.62mm/0.3in	—
L42A1	United Kingdom	7.62mm/0.3in	750m/820yds
M40A1	United States	7.62mm/0.3in	800m/875yds
M14	United States	7.62mm/0.3in	800m/875yds
M21	United States	7.62mm/0.3in	690m/750yds
Dragunov SVD	Soviet Union	7.62mm/0.3in	1300m/1422
Accuracy Int. L96	United Kingdom	7.62mm/0.3in	1094m/1000
M82A1 Barrett .50	United States	12.7mm/0.5in	1830m/2000
Heckler & Koch SG-1	Germany	7.62mm/0.3in	400m/437yd
Heckler & Koch PSG-1	Germany	7.62mm/0.3in	800m/820yd
Mk 12 Special Purpose Rifle	United States	5.56mm/0.223in	550m/600yc
Accuracy Int. L115A3	United Kingdom	8.59mm/0.338in	1400m/153
McMillan TAC-50	United States	12.7mm/0.5in	2000m/219

GLOSSARY

Anti-materiel: A role in which a sniper is tasked with attacking enemy equipment rather than personnel, such as radar or radio systems, vehicles and other military hardware.

Ballistics: A body of science connected with the behaviour of projectiles. *Internal Ballistics* is concerned with conditions inside the weapon as it is fired; *External Ballistics* is concerned with the projectile in free flight; *Terminal Ballistics* deals with the behaviour of the projectile once it has struck the target.

Bolt-action: A bolt-action weapon may be a single-shot design, loaded directly into the chamber, or fed from a magazine. Either way, the spent case is not ejected until the bolt is manually worked. This makes it easy to collect spent cartridge cases but does not allow for very rapid shooting.

Bullet drop: The effect of gravity on a bullet in flight will cause it to drop. Over a short distance this is insignificant but with a long shot, bullet drop must be compensated for or the round will fall short.

Concealment: Concealment is any obstacle or object that will obscure sight but provides little protection from a bullet or shell fragment to anyone taking cover behind it.

Cover: Cover, or 'hard cover', is any obstacle or object that will protect someone behind it from a bullet or shell fragment.

Designated marksman: An infantry soldier trained to a high standard of accurate shooting and equipped with a precision rifle, but who lacks the advanced stealth and concealment skills of a fully trained sniper. Designated marksmen generally operate with and as part of an infantry force.

Flash hider: A short extension to the barrel of a weapon which conceals the 'flash' of burning muzzle gases as they leave the barrel. A flash hider makes it much harder for hostiles to visually pinpoint a sniper's location by observing his muzzle flash, and also aids in night shooting by preserving the sniper's night vision.

Ghillie suit: A form of camouflage invented by nineteenth century Scottish gamekeepers known as 'Ghillies'. A ghillie suit is formed of cloth strips fixed to a camouflage coverall, concealing the characteristic outline of the human body.

Hold off: The distance between the aim point and the target, as seen through the sniper's sights. Hold off is necessary to compensate for bullet drop, wind effects or to lead a moving target.

Lead: It is necessary to 'lead' a moving target, to compensate for the time taken for the bullet to reach it. Greater lead is required for a fast-moving or distant target.

Lock time: The delay between squeezing the trigger and the weapon actually discharging, a term derived from the firing mechanism of early firearms, known as a 'lock' – e.g., flintlock, matchlock, etc.

Marksman: A skilled and accurate shooter, or a holder of a formal shooting qualification. Many police snipers could be more correctly termed marksmen as they lack the stealth and concealment skills of a military sniper, but by convention they are usually termed snipers.

Muzzle brake: A device that redirects some of the muzzle gases generated by firing a weapon in a direction which counteracts recoil and the tendency of the muzzle to rise. A muzzle brake makes a powerful weapon much more controllable.

Muzzle energy: The kinetic energy of a bullet as it leaves the weapon. Kinetic energy is a function of the velocity and mass of the bullet: greater muzzle energy equates to a shorter flight time, a flatter ballistic trajectory and greater wounding potential.

Observer: A sniper's assistant, who undertakes specific tasks when the team is in the field and who sometimes doubles up in the 'shooter' role. Many observers are inexperienced snipers receiving in-the-field instruction, but some are long-term professional observers.

Rifling: Spiral grooves cut into the barrel of a weapon to spin the bullet as it passes. A spun bullet is gyroscopically stabilised and therefore much more accurate than otherwise. Any longarm that possesses a rifled barrel is technically a 'rifle'.

Semi-automatic: Also referred to as self-loading, a semi-automatic weapon uses the energy of firing a round to eject the spent case and chamber the next. This may not always be desirable for a sniper, as spent cases may land outside his cover or attract attention as they reflect light. The internal workings of the weapon can also disrupt the aim point.

Smoothbore: A non-rifled weapon such as a musket or shotgun, which fires a ball or group of pellets without imparting spin stabilisation. Smoothbore weapons are inherently inaccurate and generally unsuitable for sniping, though a long barrel can somewhat compensate.

Sniper: The leader and main shooter of a sniper team, or a graduate of a formal sniper training school. Also, a military marksman trained to observe and shoot from a concealed position.

Suppressor: A device designed to reduce the noise of a weapon being fired by trapping some of the muzzle gases. It is not possible to completely 'silence' a weapon; there will always be some sound upon firing.

Tick suit: A camouflage suit made from material that prevents infrared (thermal) radiation from passing through it, concealing the sniper from thermal cameras.

Trajectory: The path followed by a bullet in flight, which is a ballistic arc caused by the interaction of the projectile's muzzle energy, gravity and air resistance.

Weld: Holding a rifle correctly, with the cheek pressed against the stock, creates a 'weld', which ensures that the relationship between eye, sight, and muzzle is always exactly the same.

INDEX

Page numbers in *italics* refer to illustrations.